4th International Symposium on the Science and Technology of Light Sources

Karlsruhe, Germany
7-10 April 1986

ISBN: 978-1-5108-4108-6

Printed from e-media with permission by:

Curran Associates, Inc.
57 Morehouse Lane
Red Hook, NY 12571

Some format issues inherent in the e-media version may also appear in this print version.

Printed by Curran Associates, Inc. (2017)

For permission requests, please contact Foundation for the Advancement of the Science & Technology of Light Sources (FAST-LS) at the address below.

Foundation for the Advancement of the Science & Technology of Light Sources
FAST-LS
Belmayne House
99 Clarkehouse Road
Sheffield, United Kingdom
S10 2LN

www.fast-ls.org

Additional copies of this publication are available from:

Curran Associates, Inc.
57 Morehouse Lane
Red Hook, NY 12571 USA
Phone: 845-758-0400
Fax: 845-758-2633
Email: curran@proceedings.com
Web: www.proceedings.com

TABLE OF CONTENTS

INVITED PAPERS

HIGH PRESSURE LAMPS WITH METAL HALIDES

HIGH PRESSURE LAMPS WITH SODIUM AND XENON

LOW PRESSURE LAMPS

INCANDESCENT LAMPS

DIAGNOSTICS

LAMP CHEMISTRY

ELECTRONIC BALLAST

MATERIALS AND MISCELLANEOUS

POST DEADLINE PAPERS

4TH INTERNATIONAL

Symposium

ON THE SCIENCE AND TECHNOLOGY OF

Light Sources

INVITED AND CONTRIBUTED

Papers

UNIVERSITY KARLSRUHE - F.R.GERMANY

APRIL 7-10.86

4TH INTERNATIONAL SYMPOSIUM
ON THE SCIENCE AND TECHNOLOGY OF
LIGHT SOURCES

7.-10.APRIL 1986

INVITED AND CONTRIBUTED PAPERS

The Symposium is organized by the
"Lichttechnisches Institut",Prof.Dr.H.-P.Popp
University of Karlsruhe
Kaiserstr.12, D 7500 Karlsruhe 1 F.R.Germany
Tel: (0)721-608 2530

and by the

"Lichttechnische Gesellschaft e.V."
Bezirksgruppe Karlsruhe

CONTENTS

INVITED PAPERS

A - H

REVIEW OF METAL HALIDE LAMPS

Alexander Dobrusskin
OSRAM, FRG, 8000 München 90

ABSTRACT

The development of Metal Halide lamps during the last 25 years shows an increasing success in the field of TV and studio lighting, the flood- and industrial lighting and recently also in general lighting applications. The different outer bulb and arc tube designs, metal halide mixtures, chemistry, and lamp ballasts in the field of the metal halide lamp technology are reviewed critically.
The most important steps in the future development of the metal halide lamps will be to obtain:

- low wattages such as 150, 70, 35 and 20 W
- different light colours such as 3000 K and 4300 K
- a colour rendition index higher than 75 and
- an increase in luminous efficacy and lamp life by
  - finding better complex compounds and better arc tube materials (such as ceramics)
  - reducing thermal losses due to coating or doping the lamp envelopes with IR-reflective or absorbing layers
  - using electronic ballasts
  - electrodeless metal halide discharges.

1. INTRODUCTION

Basically, metal halide lamps (MH) are high-pressure mercury discharge lamps containing metal halides as additives. In general, metal halides have higher vapour pressures than the pure metals. The metal halides dissociate in the core of the discharge and the metals can be excited to emit a light whose intensity depends on the vapour pressure of the metal halide. The emitted light fills the gaps in the mercury spectrum, increasing the luminous efficacy and improving the colour rendering properties of the mercury discharge.

The first lamps have been developed in the sixties; at that time the iodides of sodium, thallium and indium were used (1...6). Later, the iodides of scandium (7...8), of different rare earth compounds such as dysprosium, holmium and thulium (9...12) as well as complex compounds with cesium and tin halide were used (13...19). Investigations and papers relating to additives have been published about nearly all the elements of the periodic table. However, there is still a tremendous potential in the combination of compounds, especially in regard to the less investigated complex compounds, as well as in the choice of the halide mixtures I, Br, Cl, F.

The requirements modern light sources have to fulfill are defined essentially by economy and design. In lamp technology, design means especially the lumen package, size and shape, and the photometric data like luminance and colour, depending on the scheduled application. For general lighting purposes the main requirements are set out in Fig. 1:

ECONOMY	* high luminous efficacy	
	* long life	* good maintenance
		* colour stability
	* small production costs	
	* small costs of the system	* ballasts and ignitors
		* fixtures
DESIGN	* small wattages of 150, 70, 35 and 20 W	
	* compact form and high luminance	
	* warm white or cool white colour of 3000 or 4000 K	
	* good colour rendition; CRI higher than 75	
	* universal burning position	
	* small and light ballasts and ignitors	
	* quick warmup and hot reignition	
	* safety	

Fig. 1: requirements metal halide lamps have to fulfill

A review of the 25 years' development of metal halide lamps shows that some of the requirements listed in Fig. 1 were quite difficult to meet, especially the colour stability, the small wattages and the universal burning position. The reasons are the insufficient thermal and chemical stability of the quartz glass arc tubes, the low vapour pressure and the high chemical activity and hygroscopy of the metal halides and the difficulty to get isothermal arc tubes if the arc is constricted and bent due to the gravity or the electromagnetic field of the leadwires. Another effect on colour and life stability is caused by the diffusion of sodium ions through the arc tube walls. We will discuss these factors later.

2. DESIGN

In the sixties, during the first years of development, metal halide lamps had been designed like the old-fashioned mercury lamps with ellipsoidal or cylindrical outer bulbs and Edison screw bases in wattages between 250 and 3500 W (11, 12). The main applications of these lamps were outdoor, industrial and sports stadia lighting (Figs. 2a, 2b).

Later on, smaller, double-ended lamps in wattages of 400, 250, 150 and 70 W were introduced (12, 16, 20) and used in commercial indoor and shop window applications (Fig. 2c). At that time, a 50 W (arc tube 30 W) metal halide plug-in lamp with an integrated ballast for base-down operation only has been tested on the US market (21). The lamp should have been a great progress in general lighting but has been withdrawn, probably because of insufficient thermal stability of the integrated ballast (Fig. 2d).

The latest development in this field have been single-ended lamps with novel single-pinch arc tubes and outer bulbs having wattages of 150, 70 and 35 W (Fig. 2e).

Compact metal halide lamps HMI for TV and studio lighting, to which I will refer in the next chapter, have a special design without outer bulb (22, 23).

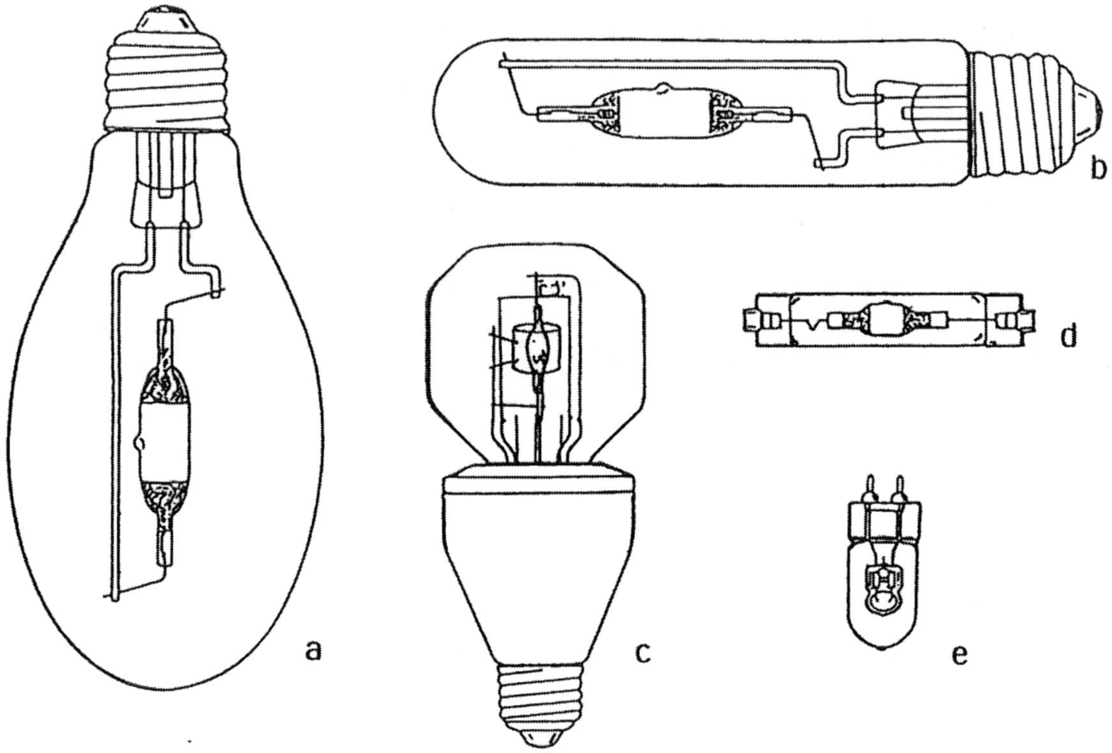

Fig. 2: outer bulb design of MH-lamps

a) ellipsoidal, E-base,

b) tubular, E-base,

c) with integrated ballast, E-base, base-down only,

d) tubular double-ended, R7s-base,

e) single-ended, G12-base

3. COMPACT METAL HALIDE LAMPS HMI

Highly-loaded metal halide lamps have been developed for TV and studio lighting. They are manufactured mainly with rare earth halides and attain daylight colour (5600 K), very good colour rendering properties (the colour rendering index is higher than 90) and high luminous efficacies (up to 102 lm/W). Outer envelopes such as high-intensity discharge lamps for general lighting required for reducing thermal losses can be dispensed with owing to the high wall loading of the arc tube. The high wall loading of such lamps, however, entails shorter lamp life such as 250 to 1000 hours (22, 23, 31).

The most important HMI lamps from 200 to 12000 W are illustrated in Fig. 3.

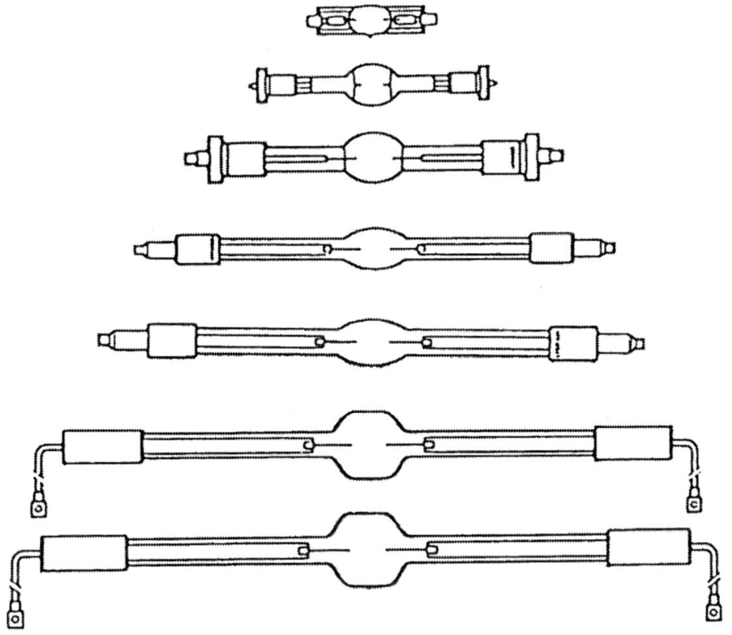

Fig. 3: HMI-lamps 200 W, 575 W, 1200 W, 2500 W, 4000 W, 6000 W and 12000 W

For the same applications,also single-ended lamps and lamps having a PAR outer bulb are on the market.

The development of the HMI lamps is continued with the following aims:

- shortening of the arc for increasing the luminance and the efficiency of the projector
- completion of the series with lamps of lower and higher wattages.

In the following I shall refer mainly to lamps for general lighting which all have outer bulbs.

4. ARC TUBES

Metal halide lamps operate with saturated vapours. There is always a condensate at the coldest spot of the arc tube which determines the vapour pressure of the halide and thus the photometric and electric data of the lamp. Arc tubes for metal halide lamps must therefore be shaped in such a way that the inner surface is as isothermal as possible. Attempts to reach this aim include giving the arc tube a special shape, applying heat reflecting coatings and using defined electrode positioning. Fig. 2b illustrates an arc tube whose shape is matched to the arched discharge. Various methods are illustrated in Fig. 4.

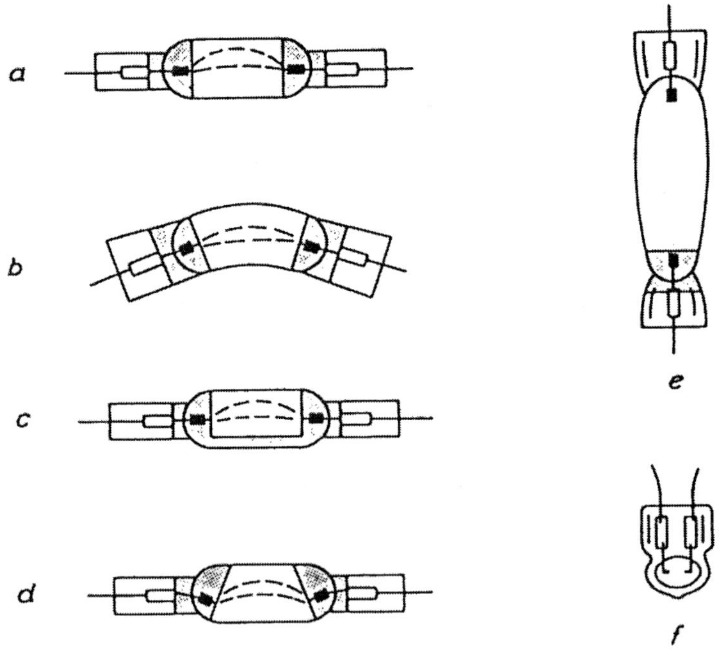

Fig. 4: design of MH-arc tubes

a) standard arc tube, any burning position,
b) bent arc tubes (Sy), horizontal burning position,
c) arc tube with additional heat reflecting coating on the bottom, horizontal burning position,
d) arc tube with downward positioned electrodes and shaped end coatings, horizontal burning position,
e) isothermal shaped arc tube, vertical burning position,
f) single-ended arc tube, any burning position

The arc tubes b, c, d and e are not advantageous, because the burning position cannot be selected freely.
In the case of the arc tubes b, c and d, special sockets and bases are required which prevent the lamp from being turned about its axis. However, with the arc tube constructions b, c and d, there can be used a higher wall loading, and higher luminous efficacies may be obtained without thermal overloading of the hottest spot of the arc tube.
Only in the cases a and f, given that suitable arc tube dimensions have been selected, nearly uniform cold spot temperatures can be obtained in all burning positions and thus also the electric and photometric lamp data independent of the burning position.

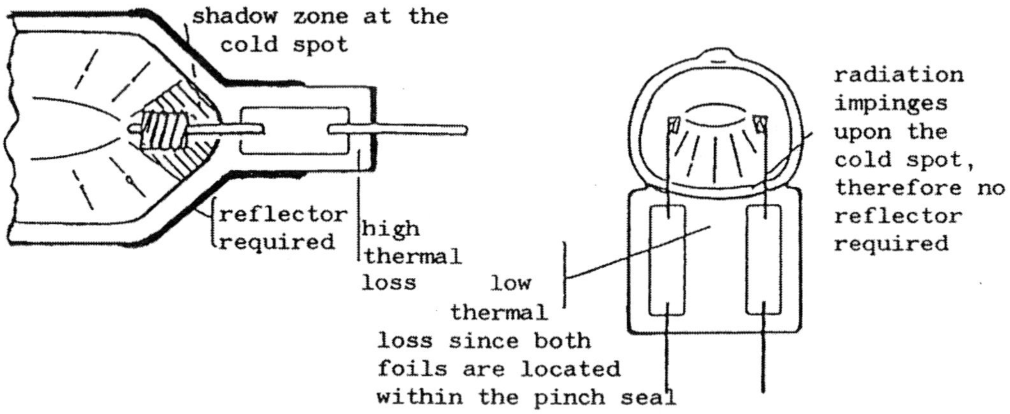

Fig. 5: comparison of double- and single-ended arc tubes

The arc tube of Fig. 5, which has a pinch seal at only one end, has the advantage that the arc radiation heats up the coldest regions with the highest heat dissipation and thereby renders the heat reflector coatings superfluous which, in the case of double-ended arc tubes, heat up the cold shadow zones behind the electrodes.

5. METAL HALIDES

In modern commercially available metal halide lamps different metal halides are used; the typical colour and efficacy data of lamps with about 70 to 250 W are given in Fig. 6.

		Na-Sc-Th	Dy-Ho-Tm-Cs-Tl	Dy-Ho-Tm-Na-Tl	Na-Sn-Tn-In-Li
CCT	K	3900	5200	4300	3000
CRI		70	93	85	76
η	lm/W	80	72	75	67

Fig. 6: metal halide additives and photometric data

The metal halides in the first column are preferred in the USA, the others in Europe. The latter have better colour properties but lower efficacies than the former.
As halogens, iodine or a mixture of iodine and bromine is used.
To get a warm colour, that means a low colour temperature of the light, combined with high efficacy, one needs sodium halide as additive.
The sodium ion Na^+ is in diameter about as small as helium and hydrogen atoms and can therefore diffuse very quickly through the quartz glass of the arc tube. Indeed one can observe such sodium loss during life especially if leadwires are passing the arc tube. The sodium loss causes instability of colour, increases the arc voltage and shortens the life of the lamp.
Waymouth et al (24) measured that photoelectrons generated on the leadwires cause a negative charge on the outside surface of the arc tube as they suck the Na^+ ions through the quartz glass.
Measurements of Nagel et al (25) yielded results which seem to contradict the model of Waymouth et al and have to be clarified. In fact, the sodium loss can be decreased by using thin leadwires spaced far apart or ceramic or glass covered wires in gas filled outer bulbs to prevent the generating of photoelectrons.

Other possibilities are to use double-ended lamps or lamps with single-ended arc tubes. See Figs. 2c and 2e. In both cases there is no need for leadwires passing the arc tube. If the outer bulb has clean inner quartz glass surfaces which do not generate photo-electrons, there will be hardly any sodium loss during life, even if the outer bulbs are evacuated.

Another possibility to prevent sodium loss is not to use any outer bulbs at all. In this case, the heat loss decreases the efficacy and colour properties or shortens the lamp life if the arc tubes are to be loaded more highly. Attempts to design lamps with warm colour and high efficacy without using sodium halide additives have not been very successful till now. Nevertheless, there are possibilities to attain CCT = 4125 K and CRI = 98 by using very highly loaded arc tubes and rare earth iodides. Fig. 7.

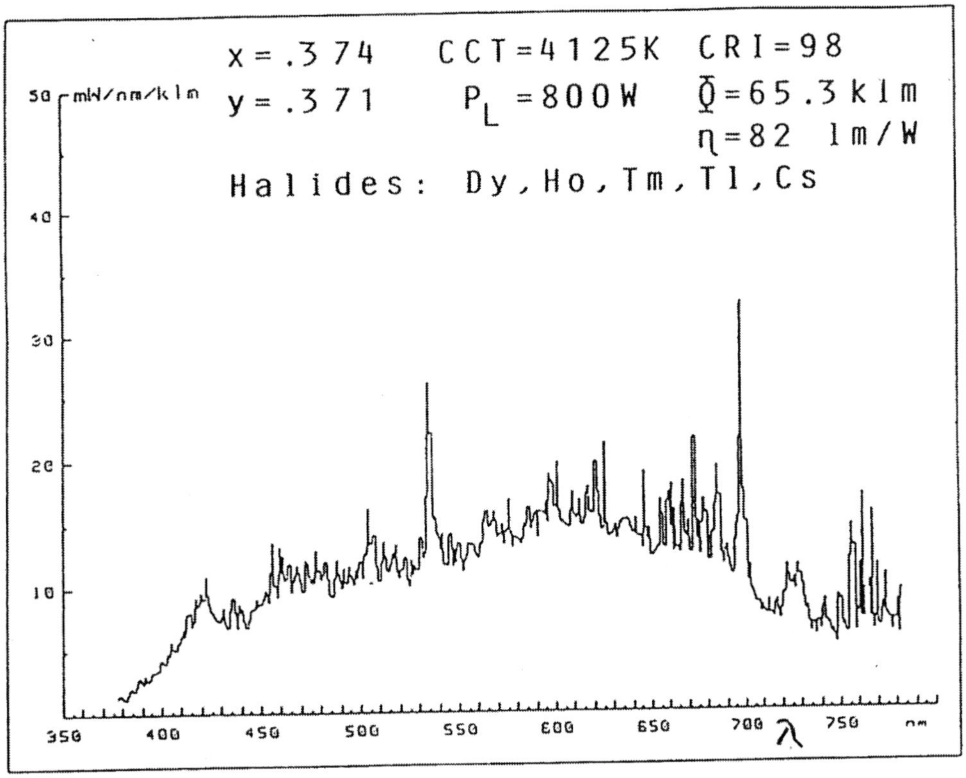

Fig. 7: 800 W-loaded MH-lamp (400 W-type) gives excellent CRI = 98 and cool white colour with CCT = 4125 K, efficacy 82 lm/W

The result of the reaction between metal halide filling and quartz glass at such high temperatures, however, again is short lamp life like the HMI compact sources have.

6. COMPLEX COMPOUNDS

The complex compounds that can be used in metal halide lamps have the characteristic that at the operating temperature of the arc tube they evaporate without becoming dissociated. By selecting desired compounds, the particle density of a desired element, for example, of dysprosium, can be increased in the arc because the complex compound has a higher vapour pressure than the dysprosium iodide alone. An example taken from the practice is illustrated in Fig. 8.

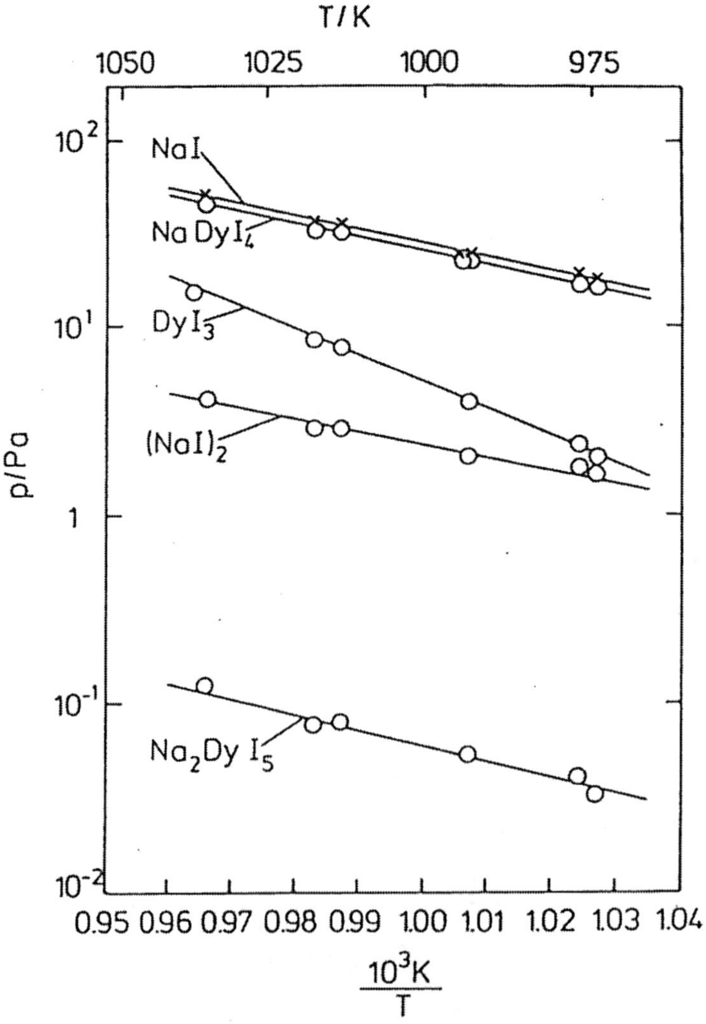

Fig. 8: partial pressures of the molecules of NaI, $NaDyI_4$, DyI_3, $(NaI)_2$ and Na_2DyI_5 above a composition of 50 mol % NaI and 50 mol % DyI3.

The sodium-dysprosium-iodide complex $NaDyI_4$ has at 1000 K a vapour pressure which is higher by a factor of about 8 than the vapour pressure of dysprosium iodide DyI_3 (37).
In the case of the sodium-tin-iodide complex, the vapour pressure at 1000 K is even raised by the factor of about 10 (26).

The investigation and use of the complex compound makes it possible to have at a relatively low wall temperature of the arc tube a relatively high particle density of such elements in the arc that have high luminous efficacies and good colour properties.

7. CHEMISTRY

Due to the large number of components in a metal halide arc tube, a large number of chemical processes occurs which, at the high temperatures inside the arc tube, are characterized not only by the materials used intentionally but also, quite significantly, by impurities. These processes may either be positive or negative with respect to the desired lamp properties. The tungsten halogen cycle which is cleaning the arc tube walls from evaporated tungsten or the formation of complex compounds, for instance, are positive examples for phenomena that may occur in metal halide lamps. The Na^+ diffusion, more a physical phenomenon, is a negative example. Apart from the fill materials that may react with the quartz glass walls and the electrodes, impurities in the form of hydrogen, oxygen, water and carbon compounds play a substantial role in the transport reactions within the arc tube. Both the life and the electrical and photometric data of the lamps depend strongly on such reactions. Utmost cleanness, defined materials and reproducible manufacturing conditions are the preconditions for the successful production of metal halide lamps (27, 28).

8. COLOUR

The colour of the light and the colour rendering properties can be selected relatively freely due to the choice of the halides. The main difficulty lies in the colour divergence from one lamp to the other and in the colour alteration during the life of the lamp. As mentioned above, the main cause for this is to be found in the cold spot temperature. The cold spot temperature depends either directly or indirectly on various factors; it controls the density of the metal halides and thus the colour of the lamp as shown by the diagram in Fig. 9. The challenge to the discharge lamp technology of today and of the future is to keep constant the factors set out in the diagram.

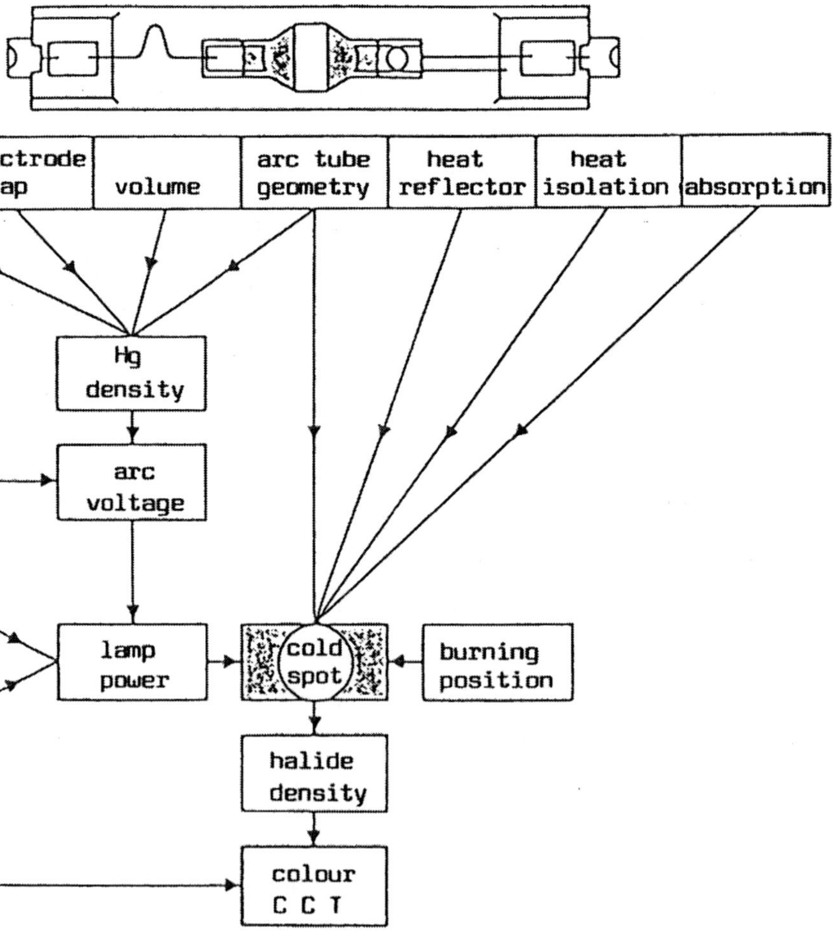

Fig. 9: influences on the colour and on the colour temperature CCT

9. LAMP BALLASTS

Conventional lamp ballasts such as chokes and transformers are still mainly used for current limiting; ignitors are often used in addition. These lamp ballasts are relatively large and heavy and constitute therefore an obstacle to the introduction of small metal halide lamps. For this reason, electronic ballasts are being developed which operate at frequencies higher than 50 or 60 Hz and have the following advantages over the conventional lamp ballasts:

- their weight and volume is lower,
- the efficacy of the system is higher,
- they include a starting device,
- their energy consumption is constant and does not depend on the supply voltage and the arc tube voltage,
- the colour divergence is thus reduced,
- flickering is reduced and the arc bending is slighter.

The difficulty in the development of such electronic ballasts lies in the selection of a suitable frequency and modulation of the supply voltage.
With many frequencies, high-pressure discharges are unstable or extinguish because acoustic resonances can be generated in the plasma of the discharge. The resonant frequency depends essentially on the size, the shape and the filling of the arc tube (32..36).

10. OUTLOOK

Apart from the above discussed possibilities, what other possibilities are there to improve the luminous efficacy and the colour rendition of metal halide lamps?
The examples of 70 W lamps - which contained different halides and were measured in the range between 20 and about 100 W/cm^2 - shows that with a higher wall loading there may be attained substantially higher luminous efficacies and better colour properties.
Fig. 10.

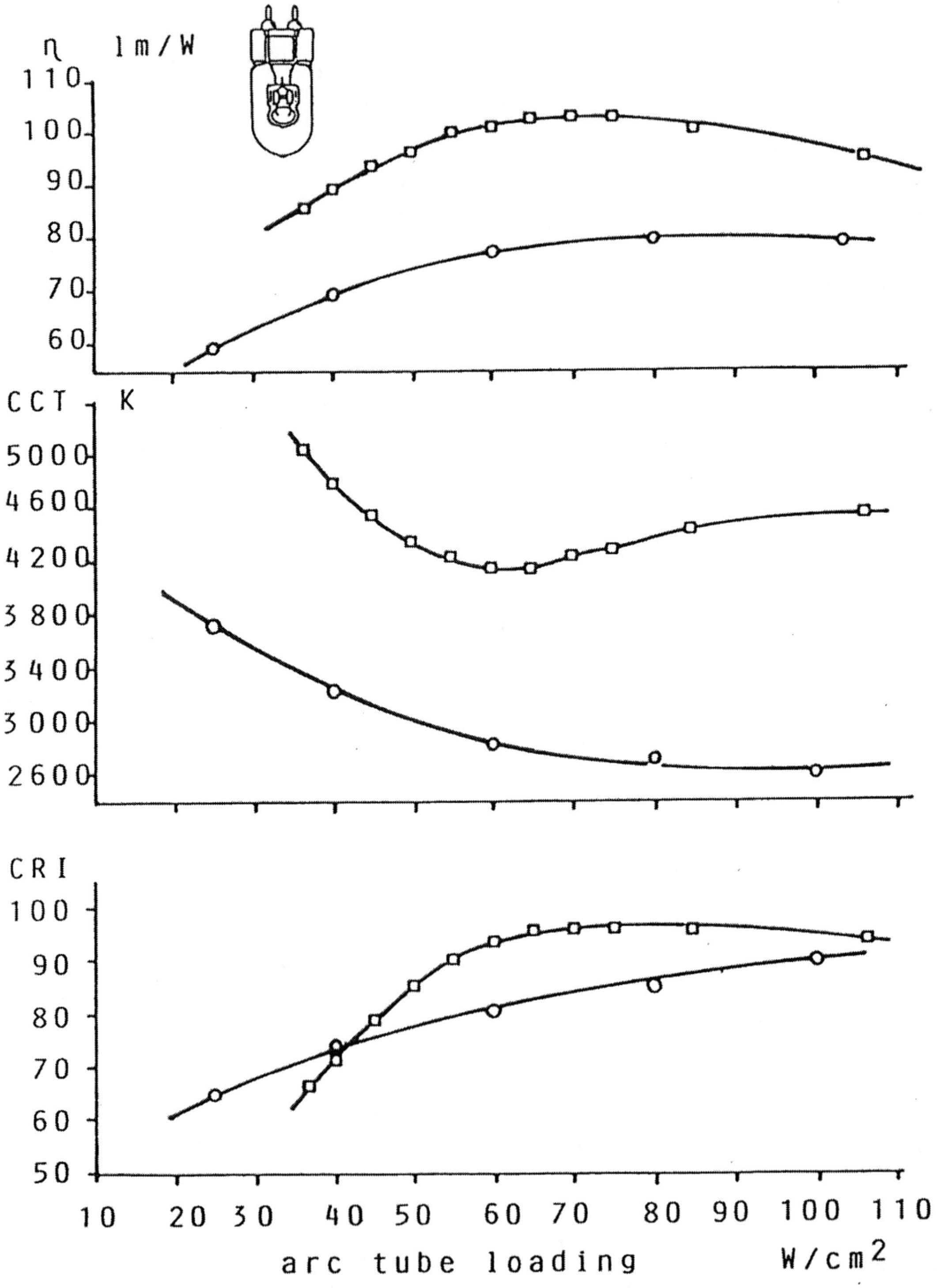

Fig. 10: efficacy η, CCT and CRI from highly loaded MH-lamps,
type □ Dy, Ho, Tm, Na, Tl-halides
type ○ Na, Sn, Tl-halides

However, the life of such lamps is considerably shortened because chemical reactions take place between the lamp filling materials and the quartz glass wall or because the softening point of the quartz glass is reached.
With sodium-scandium halides and increased wall loading, the 100 lm/W limit can be easily exceeded too but in this case the colour rendering index will only rise to 71.

These examples show that the development of the metal halide lamps has to be continued with the aim of finding materials that can withstand higher thermal strain and have a better resistance to chemicals than quartz glass.

Alumina ceramic, the material used for high-pressure sodium discharge lamps, could constitute the solution to the problem. Sufficient resistance against chemical reactions, however, is not yet ensured with the conventional sealing technology that uses niobium inleads and glass solder. Other developments (29) using cermet inleads have not yet been concluded successfully either.

The most important steps in the future development of the metal halide lamps will be to obtain:

- low wattages such as 150, 70, 35 and 20 W
- different light colours such as 3000 K and 4300 K
- a colour rendition index higher than 75 and
- an increase in luminous efficacy and lamp life by
  - finding better complex compounds and better arc tube materials (such as ceramics)
  - reducing thermal losses due to coating or doping the lamp envelopes with IR-reflective or absorbing layers
  - using electronic ballasts
  - electrodeless metal halide discharges (30).

11. REFERENCES

1. B. Kühl, H. Krense, DBP 1 184 008 (12.8.1960)
2. D.A. Larson, H.D. Fraser et al, Illum. Engng. 58 (1963), no.6, pp 434 - 439
3. E.C. Martt, L.J. Smialek et al, Illum. Engng. 59 (1964), no.1, pp 34 - 42
4. G.H. Reiling, J. Opt. Soc. of America 54 (1964), no. 4, pp 532 - 540
5. B. Kühl, Lichttechnik 16 (1964), no. 4, pp 68 - 71
6. L.B. Beijer, C.A.J. Jacobs, T. Tol, Philips Techn. Rdsch. 29 (1968), no. 11/12, pp 345 - 354
7. A. Bauer, Lichttechnik 16 (1964), no. 3, pp 118 - 120
8. Koury and Waymouth, U.S. Patent no. 3,407,327 (1968)
9. D.A. Larson, M.C. Unglert, IEEE-IGA Conference, Chicago, Ill. Oct. 1965
10. F.Rokosz, W.J. Decker, H.D. Fraser, National Technical Conference of the Illuminating Engineering Society, Montreal, Canada, Sept. 1967
11. A.Dobrusskin, Lighting Research and Technology, vol. 3, 1971, no. 2, pp 125 - 132
12. A. Dobrusskin, H. Leyendecker, Technisch-Wissenschaftliche Abhandlungen der OSRAM-Gesellschaft, vol. 11, 1972, pp 163 - 188
13. C.S. Liu, R.J. Zollweg, Journal of Chemical Physics, vol. 60, 1974, no. 6, pp 2384 - 2390
14. A.G.Chalmers, D.O. Wharmby, F.L. Whittaker, Lighting Research and Technology, vol. 7, 1975, no. 1, pp 11 - 18
15. R. Lorentz, International Symposium on Incoherent Light Sources, Loughborough, April 1975
16. A. Dobrusskin, Lichttechnische Gemeinschaftstagung Mannheim, 1984
17. D.C. Fromm, J. Seehawer, J.W. Wagner, B. Winzerling, German Patent Appln. no. 26 55 167.7, filed Dec. 6, 1976
18. D.C. Fromm, J. Seehawer, J.W. Wagner, Lighting Research and Technology, vol. 11 (1979), no. 1
19. H. Lange, D.C. Fromm, BMFT-Vorhaben ET 5062 A, 1981

20. A. Dobrusskin, D.C. Fromm, J. Heider, Technisch-Wissenschaftliche Abhandlungen der OSRAM-Gesellschaft, vol. 12 (1986), pp 11 - 30

21. W.H. Lake, J.M. Davenport, Journ. of the Ill. Eng. Soc. 11 (1982) 2, pp 60 - 73

22. A. Dobrusskin, H. Leyendecker, Fernseh- und Kinotechnik, 24 (1970) H. 7, pp 222 - 226

23. W. Block, Technisch-Wissenschaftliche Abhandlungen der OSRAM-Gesellschaft, vol. 11, pp 189 - 197

24. J. F. Waymouth et al, Illum. Engng., vol. 62 (1967), p.214

25. F. Nagel et al, High Temperature Chemistry, vol. 85-2, p.72 (1985)

26. K. Hilpert, J. Seehawer, Technisch-Wissenschaftliche Abhandlungen der OSRAM-Gesellschaft, vol. 12, 1986

27. D.E. Work, J. Electrochem. Soc., 127 (1980) 8

28. J.F. Waymouth, High Temperature Chemistry, vol. 85-2, p.1 (1985)

29. K.E. Brown, A.G. Chalmers, D.O. Wharmby, J. of IEC (1982) pp 106 - 114

30. Kramer et al, Journ. of the Electrochem. Soc. 124 (1977) 8, p. 348

31. B. Lewandowski, Technisch-Wissenschaftliche Abhandlungen der OSRAM-Gesellschaft, vol. 12 (1986), p.83

32. H. Witting, J. Appl. Phys. 49 (1978) 2680

33. R. Zollweg, J. of IEC, Jan. (1979) 90

34. W.H. Lake, J. M. Davenport, Journ. of IES (1982) p.66

35. J.M. Davenport, R.J. Petty, J. of the Illumin. Engng. Soc. 12 (1985) no. 4, pp 633 - 642

36. E. Statnic, Technisch-Wissenschaftliche Abhandlungen der OSRAM-Gesellschaft, vol. 12 (1986), p.394

37. K. Hilpert, this conference

NEW LAMP COATING AND MATERIALS MODELING

Thomas Soules and Gary Weber
General Electric Company
NELA Park, Cleveland, Ohio 44112

I. FLUORESCENT LAMP COATINGS

In recent years there has been a rapidly increasing demand for high performance fluorescent lamps: lamps with high efficiency, good maintenance, superior color rendering ability, and at low cost. This demand is being met by more sophisticated and optimized lamp coatings.

A. COATING CONSTRUCTION

Figure 1 shows a schematic of a present day multiple layer fluorescent lamp coating.

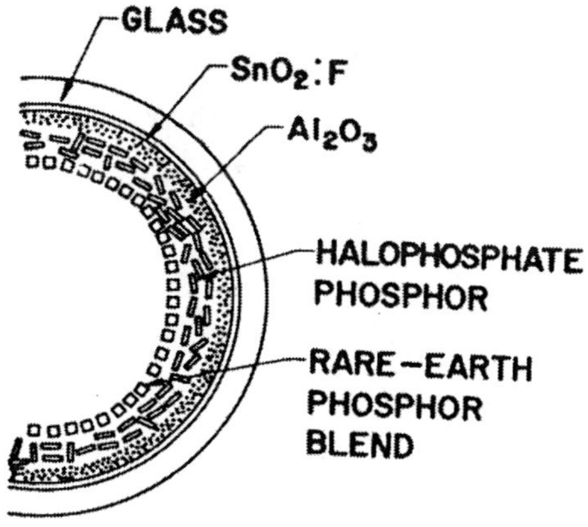

Figure 1. F-lamp Construction

To facilitate starting in lamps using neon, krypton and argon buffer gases which give a higher electric field than conventional mercury argon mixtures alone, the glass tubing is first coated with a transparent semiconductor, SnO_2:F, applied by spraying an alcohol solution containing the halides on the heated glass tubing. To improve lamp maintenance, a fine particulate alumina is applied over the semiconductor coating.

The phosphor coating consists of two layers: the outermost layer being a standard calcium halophosphate phosphor applied by downflushing a water based suspension containing the phosphor and a polyacrylic acid binder. After drying the lamp is coated with a blend of three rare-earth activated phophors in either an organic or a water suspension. More recently experiments at G.E. with electrostatic coatings of these phosphors have been successful. Electrostatic coating, which has long been used in other applications, eliminates the need for solvent removal and provides the least chance of contamination of the phosphor coating.

Because the inner layer is first exposed to the uv radiation from the discharge, the double layer phosphor construction allows much more efficient use of the rare-earth phosphor coating. For example, if the rare-earth phosphor particles are opaque to uv radiation then the fraction of uv absorbed by the innermost layer will equal the fraction of surface area covered. However, absorption of all of the uv radiation in the rare-earth phosphor layer, because of the random nature of the coating deposition, would require at least three layers of phosphor particles.

B. OPTIMUM SPECTRAL POWER DISTRIBUTIONS

The design of these phosphor systems is carried out by first simulating the lamp performance. Lamps coated with individual phosphors are made and evaluated as to color, lumen output, etc. This information is stored in files. The individual spectral power distributions are then "blended", added together, in a computer program to achieve a particular lamp color, usually, cool white, 4200K, white, 3500K, or warm white, 3000K, although other colors are also used. The underlying halophosphate spectrum is added in increasing amounts corresponding to successively thinner coatings of the rare-earth phosphor blend. The fraction of each phosphor is calculated along with projected performance of the coating: luminous efficiency, lumen output and color rendering indices.

G.E. is now marketing several new series of fluoresent lamps. Lamps labeled SP with a R_a of $\approx$ 70 have around .5 of a monolayer of a rare-earth phosphor blend consisting of the red phosphor, $Y_2O_3{:}Eu^{3+}$, the green phosphor, either $Ce_{.65}Tb_{.35}MgAl_{11}O_{19}$ or $Ce_xTb_yLa_{1-x-y}PO_4$ and the blue phosphor, $BaMg_2Al_{16}O_{27}{:}Eu^{2+}$ coated over a standard calcium halophosphate phosphor. A second line labeled SPX with still higher color rendering, $R_a \simeq 80$, uses an inner coating of about .75 of a monolayer of the same rare-earth phosphor blend. A very high color rendering lamp, R_a = .93, is achieved with a .6 monolayer of a blend of phosphors including the phosphor $Gd_{1-x}Ce_xMgB_5O_{10}:Mn^{2+}$ coated over the halophosphate phosphor. The new very high color rendering phosphor system contains all broad band emitting phosphors and is designed so that the inner coating layer supplements the halophosphate spectrum.

C. DESIGN TO MINIMIZE PHOSPHOR COATING WEIGHT

To optimize the light absorbing and scattering characteristics of the coating, Monte-Carlo simulations of the uv scattering were performed. The individual phosphor particles are represented either as spheres or as spherical in shape but with randomly oriented planar surfaces.

Figure 2. Models for Monte-Carlo Light Scattering Calculations

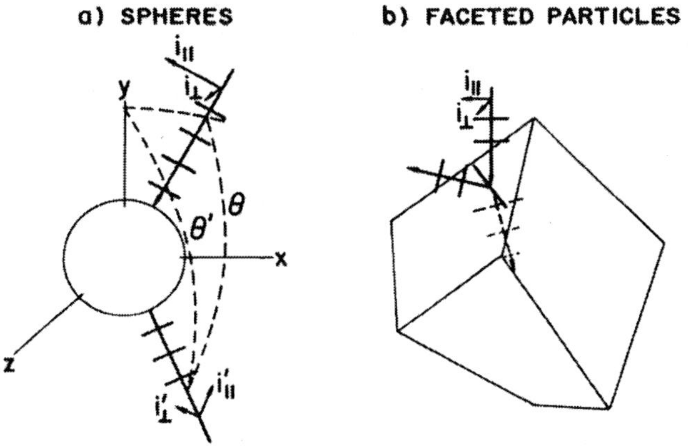

In the case of spherical particles, after each encounter with a particle the direction of the scattered ray is determined using the Mie scattering solutions and the intensity of the ray is modified by the fraction of the uv which is absorbed. Ultraviolet rays are traced from one particle to another. In the case where the particles are faceted, Kettler's laws are used to determine the direction and intensity of the reflected and refracted rays from a planar surface. Rays are then traced from one surface to another and intensity is reduced by attenuation within the absorbing medium.

In order to determine the absorption coefficient of the phosphor, the percent uv reflected off a thick plaque is measured and compared with simulations such as shown in Figure 3.

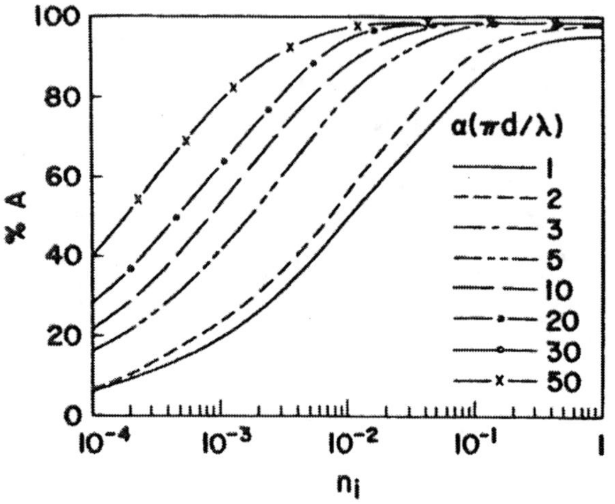

Figure 3. UV absorbed by a thick plaque of spheres (n_r=2.0)

Complex indices of refraction so obtained are used together with simulations for coatings of finite thickness. These indices determine if the phosphor's absorption is sufficiently high for a coating and what minimum particle size will result in a coating which will absorb some fraction of the uv with the smallest possible number of particle layers. The use of larger particle sizes for the moderate and strong absorbers does not change the number of layers needed to absorb a given fraction of the uv. From these results, we have determined that an optimum minimum particle size for our current yttrium oxide phosphor is around 5 mic and for the green phosphor it is around 1-2 mic.

Further investigations have been made with different coating covering statistics which show that still less phosphor can be used by using well-dispersed coatings, i.e. inner coatings with few particles stacked on top of each other.

II. INCANDESCENT LAMP COATINGS

There are myriad coatings used on different types of incandescent lamps. Examples include lamps with teflon coatings to prevent shattering. The teflon is applied electrostatically and then melted. Decorative lamps are sold with exterior glass frits melted on the bulb. Glass frit containing neodynium is used to absorb a narrow band in the yellow green region of the spectrum so as to move the color of the lamp toward higher color temperatures. Organic laquers, acrylics and silicones are also used for sign decorative lamps and on the cover of color beam lamps with parabolic reflectors. Iridescent coatings of titania and dichroic filters are also used principally to give decorative effects. Dichroic reflectors designed to transmit heat out of the back of a reflector lamp while reflecting forward the visible light beam are widely used in projection lamps.

A. LIGHT DIFFUSING COATINGS

In the United States nearly all incandescent lamps use some means of diffusing the light of the filament. Prior to seven years ago, most of the bulbs were flushed twice with a hydrofluoric acid solution. The etched glass surface bends light rays by relatively small angles. Much better light diffusion is obtained by coating the inside of the bulb with a silica or alumina "smoke" by burning organo-metallic compounds, such as, tetraethyl silicate inside the bulb in a clean oxygen flame. The deposited silica or alumina consists of spheres with a mean diameter of .1 mic. The optical properties of such coatings were modeled using the same Monte-Carlo scheme as described above. Calculations are in good agreement with the optical properties of these coatings. Common practice, today, is to use a mixture of powders: a finely divided very white kaolin clay, a fumed silica and a fumed and aerated silica. These powders are maintained in a fluidized bed and electrostatically coated on the inside of the bulb. Reflectance values of 25 percent give acceptable light diffusion.

B. INFRARED REFLECTING FILMS

Since a filament at 2800K radiates far more infrared radiation (ir) than visible light, lamp maufacturers have vigorously searched for coatings which would return ir to the filament. The first ir reflecting films were transparent semiconductors, such as, In_2O_3:Sn. These can be designed to have a plasma frequency cutoff between the infrared and visible region of the spectrum. Calculations using a simple Drude model show that such materials must have 10^{+21} carriers/cm^3 and resistivities of 10^{-4} ohm.cm and no visible or near visible interband transitions. However, the stingent criteria for a sharp reflectivity cutoff and very good visible transmission were not met. The next generation ir reflecting coating consisted of a very thin silver film with a ZnS interference filter to minimize the silver reflectivity in the visible. Currently interference filters consisting of alternate layers of silica and tantala are being used on certain incandescent halogen lamps at G.E. These coatings offer the advantage of high temperature stability, little or no visible absorption and considerable flexibility in the design of the reflectors wavelength response. An example of a reflectance curve is shown in Figure 4.

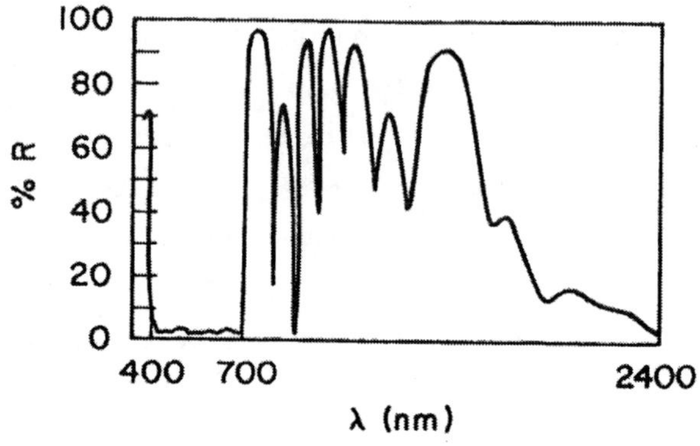

Figure 4. Reflectance curve for SiO_2/Ta_2O_5 ir reflecting film.

To optimize the efficiency of the lamp, both the film and lamp geometry must be modeled. A schematic diagram of lamp is shown in Figure 5.

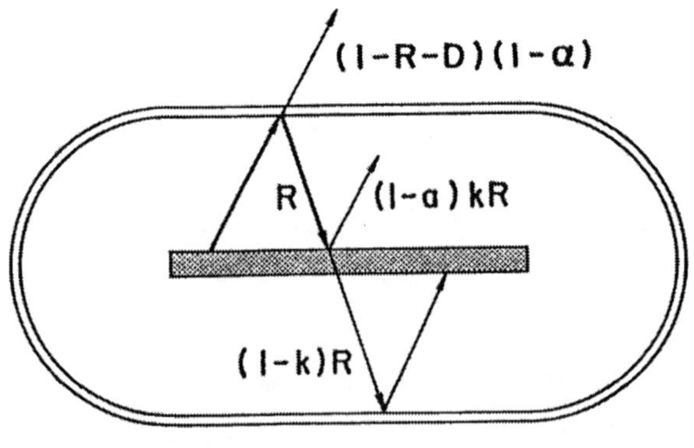

Figure 5. Schematic of tungsten-halogen lamp with ir reflecting film.

The fraction of ir transmitted through the coating is (1 - R - D) * (1- a) where R is the fraction specularly reflected and D the fraction diffusely reflected. a is the fraction absorbed by the film. A fraction of the specularly reflected ir strikes the filament and is absorbed and a fraction is rereflected by the filament. Specularly reflected ir which misses the filament may be imaged back on the filament after a second reflection if the filament is slightly off-axis. End losses are very important as is high reflectivity of the coating in the infrared and low reflectivity in the visible. The coating and lamp system are optimized to give the maximum ir reabsorbed by the filament and the maximum visible light transmitted. Current coatings improve efficiency of halogen lamps by around 30 percent and higher values are possible. A problem with the tantala-silica filters is that high stresses occur in the film during lamp operation because of a large mismatch in the thermal expansion coefficients of tantala and the alumino-silicate glass or vitreous silica lamp envelope material.

III. OTHER LAMP COATINGS

Still other lamp coatings are applied to high intensity discharge lamps to retard sodium loss. Sodium ions can diffuse through vitreous silica at lamp operating temperatures and can also diffuse along grain boundaries and through seal glasses in lamps using a polycrystalline alumina envelope.

IV. LAMP MATERIAL MODELING

In order to minimize lamp failure caused by large stresses in glass, general finite-element and finite difference models have been developed to calculate the residual thermal stresses in glass parts. Both the specific volume and the shear stress modulus relax in the glass transition region. These relaxation characteristics of the glass are included. The programs take as input the part geometry and temperature distribution as a function of time ,as well as, material parameters. Stresses are calculated both at room temperature after forming and at the operating temperature of the lamp.

In addition, fundamental calculations relating glass properties to microstructure and composition have been carried out using the technique of molecular dynamics. In molecular dynamic calculations atoms are assigned some form of interatomic potential and a system of several hundred to several thousand atoms is followed dynamically in time. Experiments can be performed on the model. Factors effecting ultimate glass strength and sodium ion diffusion are illustrated.

RECENT PROGRESS IN LOW PRESSURE HG-RARE GAS DISCHARGE RESEARCH

Jakob Maya
GTE Lighting Products
Danvers, MA 01923

In this paper we will give an overview of recent progress accomplished in the area of low pressure Hg-Rare Gas lamp research. The topics we will cover include recent improvements in Hg-Rare Gas discharge modelling, efficiency improvements due to isotope alteration of Hg, hyperfine structure measurements of the 253.7nm Hg resonance line, radiation transfer in low pressure plasma's, Monte Carlo calculations, energy transfer between Hg isotopes, radial excited state distributions of isotopes in discharges, RF probe measurements in plasma's, and novel laser and fiber optics diagnostic techniques for characterizing low pressure discharges. The work to be presented has primarily been done at our laboratories and represents a portion of our activities over the last several years.

Considerable amount of work has been done over the years to explain the general behavior of the low pressure Hg-Rare Gas discharge which constitutes the basis of the fluorescent lamp. Recent work by the Philips Research Groups recognized the importance of including effects of non-Maxwellian electron energy distribution, associative ionization and volume recombination. However, these effects still do not seem to explain the behavior of the discharge at very low (~10-20°C) and very high (~60-80°C) mercury cold spot temperatures. In order to explain the behavior at these extreme temperatures as well as in the intermediate region we developed a model based on the equations of population, ionization and energy balance averaged over the tube cross section. The electron energy distribution function (eedf) has been obtained from an approximate solution to the Boltzman equation. This expression takes into account the depletion of the distribution tail due to inelastic collisions which are dependent on Hg and (e) number densities. This (eedf) appears to be somewhat more realistic than the 2 electron group model utilized in earlier works. About 12 types of kinetic and collisional processes each of which contain 1-6 state selective collisions have been taken into account to describe the discharge. Elementary constants for these processes have either been obtained from the literature or have been adjusted by fitting calculational results to experimental data of Hg UV output and electric field intensity in a T/12 tube.

We have recently reported on the increased efficiency of a fluorescent lamp by altering the isotopic composition of Hg. More specifically we demonstrated that by increasing the concentration of ^{196}Hg isotope from 0.146% to a few percent we offer an easier channel for the photons to escape the tube, thereby reduce the probability of radiationless transitions and increase the efficiency by as much as 4-5%. In order to further quantify the phenomenon we undertook a series of total UV and Fabry Perot hyperfine structure measurements of discharges with natural as well as isotopically altered Hg. While the total UV measurements were in good agreement with total integrating sphere visible light output measurements the hyperfine structure measurements confirmed our understanding of why we obtain the maximum efficiency increase with only a few percent ^{196}Hg and we reach saturation as we further increase the ^{196}Hg concentration. Due to very strong energy transfer between Hg isotopes, (σ=1100 $\AA^2$) of the type $^{N}Hg(6^3P_1) + {}^{196}Hg(6^1S_o) \rightleftarrows {}^{N}Hg(6^1S_o) + {}^{196}Hg(6^3P_1)$ where ^{N}Hg represents all isotopes except ^{196}Hg, just a small addition of ^{196}Hg offers an easier exit channel for some of the radiation of the other isotopes. However as we keep increasing the ^{196}Hg concentration the back energy transfer rate becomes appreciable. Furthermore the emission of ^{196}Hg is not anymore a distinct sixth channel of the hyperfine structure but rather starts overlapping with the emission of ^{199}Hg, ^{201}Hg where it can be absorbed by these isotopes. These two effects results in an optimum ^{196}Hg concentration of 2-4%. It should be noted that the efficiency improvement obtained is very close to what the ideal composition requires as determined by a regression analysis. This again is due to the effects described above.

In our search for further understanding of the low pressure discharge we ask overselves whether all Hg isotopes have the same excited state radial distributions. In order to obtain the radial distribution of excited states for each isotope we solved the Holstein equation numerically. We assumed an excitation function of the form

$$\alpha_i(x)=G_i x F_i x(1-Ax^2)\exp(-Bx^2)$$

where G_i and F_i are statistical weight and isotope fraction respectively and obtained the constants by fitting the form to known experimental data. Calculations done on a single isotope with various optical depths indicate that near the wall the distribution is less steep for lower optical thickness. When the same calculations were run for natural Hg all isotopes except ^{196}Hg had fairly similar spatial distributions. Emission from isotope ^{196}Hg however for r/R>0.8 (where R is tube radius and r is radial position) showed a less steep decline in agreement with optical depth calculations above. The key conclusion of this type of calculations is that in a multisotopic mixture if all isotopes are approximately in similar concentrations (within a factor of 2) their radial excited state distributions are fairly close. However, if one of the isotopes has a much smaller concentration its radial excited state shows a substantially flatter distribution especially near the walls (r/R > 0.7-0.8) of the discharge.

Recently we have initiated several forms of diagnostics for the low pressure Hg-Rare Gas discharge. In the following I summarize these techniques and give a few examples of the kind of data we obtained.

The first technique we describe is an RF double probe measurement of the Langmuir type. This technique has been developed in our laboratories by Dr. Valery Godyak and you will hear more about it in a later presentation. These measurements have been performed using a new electronic circuit which overcomes most of the difficulties encountered in Hg-rare gas plasma measurements, such as plasma noise, change of probe work function due to Hg amalgamation etc. The technique exploits a combination of fast pulsing with a second probe in close proximity to the main probe for noise feedback. Using high speed operational amplifiers and a computerized data acquisition system highly accurate data on (eedf), plasma density and mean electron energy is obtained. A series of measurements in Hg-Ar at various pressures of rare gas and Hg indicated a significant deviation from a Maxwellian distribution. These measurements are in good agreement with our calculations of (eedf) using the model described above. Examples of data will be discussed.

We have been developing an intersecting laser beam technique which has the potential of yielding point information about the radial distribution of excited states in the positive column. This technique primarily developed by Dr. Philip Moskowitz of our laboratories utilizes a strong pump and weak probe beam orginating from the same laser and intersecting at right angles at a point in the discharge. The saturating pump beam is chopped and modulates the population of the particular excited state. Using phase sensitive detection on the probe beam one obtains highly localized excited state number density without having to resort to Abel or other kind of inversion procedures inherent in line cord radiation techniques (e.g. laser absorption, laser induced fluorescence etc.). The technique relies on the fact that when a state is saturated it is optically transparent. It also relies on the expectation that a strong saturating narrow beam does not affect appreciably the distribution of excited states far away from its path. Precautions and experiments to verify these expectations will be discussed in the full text.

Finally a third technique which we have been developing is the use of a quartz fiber optic external to the lamp coupled to a Fabry-Perot interferometer. This techniques has primarily been developed by Dr.'s Mark Grossman and Radomir Lagushenko. Here the quartz fiber optic which samples 253.7nm radiation from the discharge along its cord is translated automatically and the hyperfine structure recorded as a function of radial position. The fiber optic provides a small acceptance angle, accurate positioning and the ability to couple the lamp measurements with a Fabry Perot interferometer which may be physically far away. In order to obtain radial excited state distributions for each isotope the experimental results have to be compared to a numerical solution of the modified Holstein-Biberman equation. This equation has been solved numerically as mentioned above. The solution of the equation gives a different (hfs) for each radial excited state distribution. By obtaining the best match to the measured (hfs) we determine the radial excited state distribution. Examples of such measurements will be given, in the full text.

PREPARATION AND USE OF HIGH PURITY METAL HALIDES IN LAMPS

by

Timothy R. Brumleve, James L. Schoolenberg and Robert S. Roller
Anderson Physics Laboratories, Inc.
406 N. Busey Ave.
Urbana, IL 61801 USA

Discharge lamps occupy a very important position in lighting, as a large fraction of all artificial light in the world is produced by such means (fluorescent, mercury arc, high- and low-pressure sodium, and metal halide lamps). Metal halide lamps, in particular, can offer substantial advantages in both efficacy and color rendition over other types of discharge lamps. In addition, the spectral output of metal halide lamps can be tailored for particular applications by appropriate choice of fill materials.

The idea of adding metal halides to arc lamps is not new: the basic principles were discovered by Steinmetz in 1911 (1). It was not until the late 1950's and early 1960's, however, that such lamps began to be developed commercially. At about this same time Anderson Physics Laboratories was developing methods for the preparation of ultra-pure metal halides (primarily chlorides) for use in color center research, display technology, and molten-salt electrochemistry. The methods of reactive gas processing and the direct synthesis from high-purity elements of the rare earth and transition metal halides were developed.

Concurrent with the development of purification and synthesis techniques, the Laboratory developed methods of manufacturing precisely-sized, spherical metal halide particles. It thus became possible to dose lamp arc tubes with a precise amount of metal halide salt, either by volumetric dosing of free-flowing spheres of small diameter, or by counting a small number of larger spheres into the arc tube.

The periodic chart shown in Figure 1 illustrates the wide range of ultra-pure, anhydrous metal halides purified or synthesized in recent years. There is, of course, an infinite number of possible bi-component and multi-component mixtures of these salts. The physical and chemical properties of such mixtures, the methods of preparing them as precisely-sized spheres, the technology for purification and synthesis of the individual components, and the analytical techniques for determining chemical composition and trace contamination are described below.

PURIFICATION AND SYNTHESIS

Metal halide materials suitable for use in discharge lamps must meet a number of stringent requirements. First and foremost is the requirement of low hydrogen and oxygen content. The effects of oxygen on electrode erosion and tungsten transport and of hydrogen on ignition and reignition voltages are well known (2,3). The most common source of O- and H-containing impurities is probably atmospheric water vapor. Other potential sources include hydroxide (OH^-), other oxyanions such as SO_4^{2-}, SiO_4^{2-}, and NO_3^-, as well as organic materials.

It is also desirable to eliminate foreign metal contamination to as great an extent as possible. Thus most materials prepared for lamp use contain less than 10 ppm (parts per million by weight) total metallic impurities.

The method of reactive gas processing (RGP) using anhydrous hydrogen halide gas is employed in the purification of the alkali metal halides such as NaI (4-6). These halides are commercially available with low metallic impurity levels, but typically contain high amounts of H_2O and traces of OH^- and other oxyanions. Referring to Figure 2, the salt is typically placed in the upper part of a fused silica vessel, and evacuated at an elevated temperature to remove free H_2O. Dry HI is then admitted through the lower port to remove hydroxide impurities:

$$HI_{(g)} + MOH \rightarrow MI + H_2O_{(g)} \qquad (1)$$

Levels of foreign halides are also reduced by halide exchange, e.g.

$$HI_{(g)} + MCl \rightarrow MI + HCl_{(g)} \qquad (2)$$

The HI treatment is continued during melting and is then replaced by purified Ar to purge the melt of any residual HI and H_2O. Finally, the melt is forced through the quartz frit and funnel to remove insoluble contaminants, and the jet of molten droplets is quenched (frozen) in a suitable gas of high thermal conductivity such as He. NaI particles prepared in this fashion typically contain less than 2 ppm of total H_2O, hydroxide and oxide as measured by coulometric Karl Fischer titration (7).

Virtually all of the other metal halides shown in Figure 1 are synthesized from the elements, the exceptions being the Group IA metals noted above and some selected chlorides. Previous approaches to the synthesis of rare earth halides have included direct elemental synthesis (4,8), reaction of HgI_2 with rare earth metals (8), and reactions involving displacement of waters of hydration by ammonium halides followed by sublimation of the ammonium halide (8-11). It is generally agreed that the highest purity halides are those synthesized from the direct reaction of the elements (8).

For example, the direct synthesis of FeI_2 has recently been described (12). It is depicted in Figure 3 in which the metal (Fe) contained in a tungsten crucible is reacted directly with

I_2 vapor from the lower portion of the reaction tube. In our efforts to improve lamp products we have shifted increasingly to direct syntheses from high-purity elements as opposed to purifying inferior starting materials. The success of this approach has been demonstrated with improved vapor pressure measurements above molten FeI_2 in the range of interest for lamp design (12,13).

The handling and packaging of purified metal halide spheres is conducted in Ar-filled glove boxes with recirculating oxygen and water removal systems (14). A good glove box should be capable of maintaining levels of less than 1 ppm (v:v) H_2O and O_2. Measurement of H_2O at the ppm level is possible by coulometric Karl Fischer titration (15) of the box gas, optical dew point measurements, thin film alumina conductivity probes (16,17), and piezoelectric based detectors (18,19).

The measurement of O_2 levels is a good indicator of overall drybox quality, since any leak or glove puncture will admit oxygen along with atmospheric moisture. We have found that the Clark-type of amperometric membrane diffusion cell (20) has a reliable detection limit of less than 1 ppm O_2.

CONTROL OF SIZE

The control of size (and hence mass) of each metal halide particle is a major concern for lamp dosing purposes. The Laboratory has developed two distinct processes for producing metal halide spheres.

Figure 4 illustrates the "dropping" technique which is useful for preparing large spheres of diameter greater than about 0.8 mm. For NaI this corresponds to masses from 9 mg to over 20 mg per particle. The system is similar to that shown in Figure 2 except that a thin-walled funnel of known outer diameter replaces the jetting nozzle. The molten halide flows through the funnel stem, and when the weight of the drop exceeds the product of the surface tension of the liquid by the circumference of the stem, the drop releases and solidifies in the condenser below. Further details can be found in the literature (21).

Figures 2 and 5 illustrate the "jetting" technique which is applied to spheres of diameter less than 0.2 mm up to about 0.6 mm. For NaI, this corresponds to masses from less than 15 μg to about 3 mg per particle. Here the liquid halide emerges from the nozzle as an unbroken stream or jet which is broken into uniform droplets by audio-frequency vibration. The frequency and jet velocity must be carefully controlled. Further details on the theory and practice of controlled jet break-up can be found in the literature(4,5).

The determination of the size and mass distribution of metal halide particles larger than about 1 mg is typically performed by optical measurement (22). Typical values for the resulting relative standard deviation (RSD) in mass fall in the range of about 5% to 8% for particles over 1 mg. The RSD in mass is relatively independent of the method ("dropping" or "jetting") employed.

SALT MIXTURES

In order to obtain the desired spectral output and operating conditions, most metal halide lamps contain at least two metal halide components (e.g., NaI-ScI_3 or NaI-TlI-InI). The feasibility of of preparing a given salt mixture depends upon the physical properties, as represented by the phase diagram, and upon the vapor pressures of each component of the mixture at the melt temperature.

As an illustration, the phase diagrams of the binary systems $NaI-ScI_3$ and $CsI-HoI_3$ are discussed (23). If the difference between liquidus and solidus lines is too great, liquid droplets may not completely quench, and solid spheres will be difficult or impossible to produce. Unfortunately, published data for most multi-component mixtures of lamp interest do not exist. Thus the Laboratory has amassed phase diagram data for such mixtures based mostly on cooling and melting curve data. This allows the selection of operating conditions for custom mixtures without exhaustive trial and error.

The vapor pressures of the components of a mixture are also critical. For example, in the preparation of the NaI-TlI-InI mixture, the individual vapor pressures at the liquidus are in the approximate ratio of 1:10:100, respectively. This can lead to the distillation of the more volatile components from the surface of the melt, and also from the surface of the liquid droplets during quenching. Compensation for these effects must be accomplished for each metal halide mixture of lamp interest.

Other physical data necessary for both the "jetting" and "dropping" procedures include the liquid and solid density, the surface tension, and the viscosity, all as a function of temperature. While these data are available for some pure metal halides, they must be measured and compiled for the wide variety of mixtures.

The advantage of dosing mixed metal halide spheres over the dosing of individual components was demonstrated by Anderson and Page (4). Improvements in color rendition and lamp-to-lamp variation are attributable primarily to the improved accuracy of the total dose and to the improved accuracy in the concentrations of the individual components.

ANALYTICAL METHODS

Chemical analysis plays a very important role in metal halide manufacture, in quality control of products, and in development of purification and processing methods for new experimental mixtures.

As noted above, levels of H- and O-containing impurities are critical to lamp performance and lifetime. Of all potential contaminants, H_2O is of most concern because of the exceedingly hygroscopic nature of most metal halide salts. For example, one pellet of purified NaI with a surface area of about 2.5 mm^2 absorbs about 10 μg per minute when exposed to dry room air (a rate of over 10 ppm/sec). All pellets containing alkali metal or rare earth halides will become grossly contaminated if mishandled.

Analytical methods which are applicable to trace O and H determination in these materials include 14-MeV neutron activation analysis for oxygen (24,29), reductive fusion for total O and H (25,26), and coulometric Karl Fischer titration for H_2O, OH^-, and oxides (7,15). Methods based on the electrical properties (2,3) and emission spectra (27,28) of completed arc tubes are also possible.

Because of its specificity for H_2O and OH^- we have turned to coulometric Karl Fischer titration (15). The application of this method to metal halide salts has recently been described (7) and was shown to respond reliably to levels of 2 μg H_2O (or to 0.2 ppm for a 10 g sample). The method does not respond to rare earth oxides or oxyiodides.

For other trace contaminants we employ the following methods. Atomic absorption and flame atomic emission spectrophotometry have virtually replaced older arc/spark optical emission and wet chemical trace metals methods in the Laboratory. When a survey analysis is required we usually employ spark source mass spectrometry (SSMS) (30). The results of a typical SSMS analysis of lamp-quality NaI are shown in Table I. Molten salt electrochemical methods are sometimes applied to the detection of H_2O and OH^- where other methods fail. Ion chromatography has been applied to the determination of trace Cl and Br in FeI_2 (12) and should be applicable to most water-soluble iodides.

The accurate analysis of the major components of mixed salts is also important, both in quality assurance and in the manufacturing process. In many cases metal concentrations are determined by atomic absorption and flame emission methods which are usually precise to about 1% RSD. In cases where these methods offer less precision (such as the determination of rare earths), wet chemical methods such as EDTA complexometric titration are employed.

In addition to these methods, we routinely employ a wide variety of classical and instrumental methods. Gravimetry, pH and ion-selective electrode measurements, argentometric, potentiometric, and redox titrations, d.c. and pulse polarography, linear sweep and cyclic voltammetry, and u.v.-visible and infrared spectrophotometry have been applied to the trace and major component analysis of metal halide lamp materials.

ACKNOWLEDGMENT

We wish to acknowledge the monumental contributions of Dr. Scott Anderson, founder and past president of Anderson Physics Laboratories, to the science and technology of metal halide lamps.

REFERENCES

1. C.P. Steinmetz, U.S. Patent No. 1,006,021 (1911).

2. J.F. Waymouth, "Electric Discharge Lamps", M.I.T. Press, Cambridge, MA, 1971.

3. M.A. Cayless and A.M. Marsden, "Lamps and Lighting", 3rd edn., E. Arnold, London, 1983.

4. S. Anderson and R.B. Page, J. Light. Visual. Env., 5, 1 (1981).

5. S. Anderson, U.S. Patent Nos. 3,676,534; 3,948,793; 4,010,000.

6. H.A. Laitinen et al., J. Electrochem. Soc., 104, 516 (1957).

7. T.R. Brumleve, Anal. Chim. Acta, 155, 79 (1983).

8. J. Burgess and J. Kijowski, Advances in Inorg. Chem. and Radiochem., 24, 57 (1981).

9. W.G.C. Verbeek and J.W. Harmsen, paper No. 64 presented at the Third International Symposium on the Science and Technology of Light Sources, Toulouse, April 1983.

10. J. Kutscher and A. Schneider, Inorg. Nucl. Chem. Lett., 7, 815 (1971).

11. M.D. Taylor and C.P. Carter, J. Inorg. Nucl. Chem., 24, 387 (1962).

12. S.A. Mucklejohn, N.W. O'Brien and T.R. Brumleve, J. Phys. Chem., 89, 2409 (1985).

13. H. Schäfer and W.J. Hönes, Z. Anorg. Allg. Chem., 288, 62 (1956).

14. Vacuum Atmospheres Co., Hawthorne, CA, USA.

15. J. Mitchell, Jr. and D.M. Smith, "Aquammetry" Part III, Wiley-Interscience, NY, 1980.

16. Reference 15, Part I, 1977.

17. Panametrics, Corp., Waltham, MA, USA.

18. J. Hlavay and G.G. Guilbault, Anal. Chem., 49, 1890 (1977).

19. G.G. Guilbault, Ion-Sel. Electrode Rev., 2, 3 (1980).

20. Delta F Corp., Woburn, MA, USA.

21. S. Anderson, U.S. Patent No. 4,201,739.

22. P.D. Lark, Anal. Chem., 39, 55 (1967).

23. A.G. Krokhina, et al., Zh. Neorg. Khim., 25, 1624 (1980).

24. S.S. Nargolwalla and E.P. Przybylowicz, "Activation Analysis with Neutron Generators", Wiley, NY, 1973.

25. W.E. Dallman in "Encyclopedia of Industrial Chemical Analysis", Snell & Ettre, eds., Vol 8, 1968, pp 613-693.

26. P. Wiedijk, Anal. Chim. Acta, 81, 105 (1976).

27. S Mochimaru et al., J. Illum. Eng. Soc., 14, 264 (1984).

28. T. Kazui et al., J. Light Vis. Env., 1, 10 (1977).

29. IRT Corp., San Diego, CA, USA.

30. Northern Analytical Laboratories, Inc., Amherst, NH, USA.

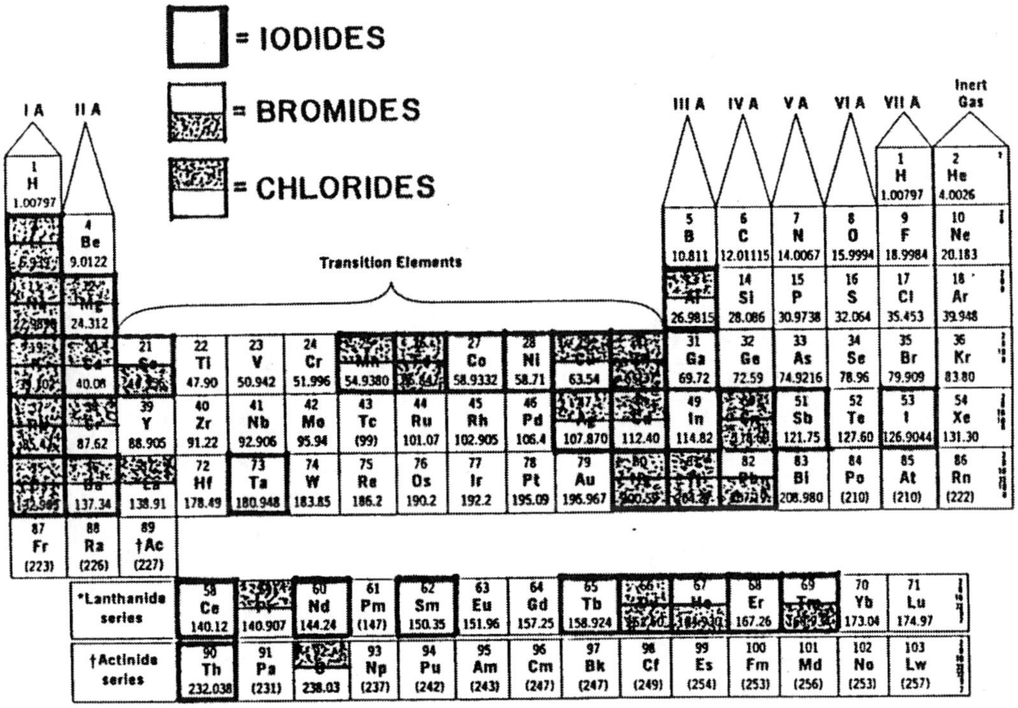

Figure 1. High-purity ultra-dry metal halides

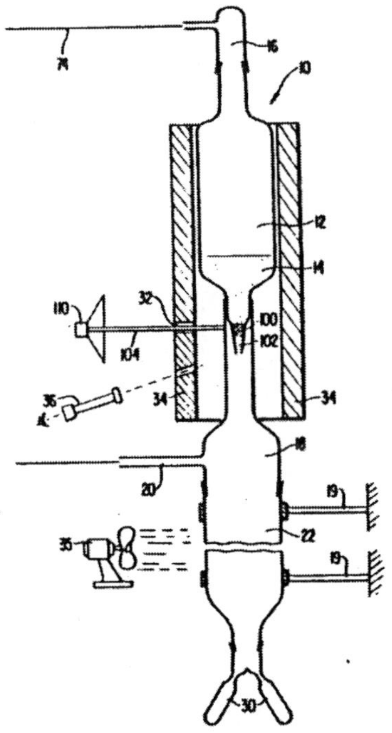

Figure 2. Purification apparatus for metal halide salts

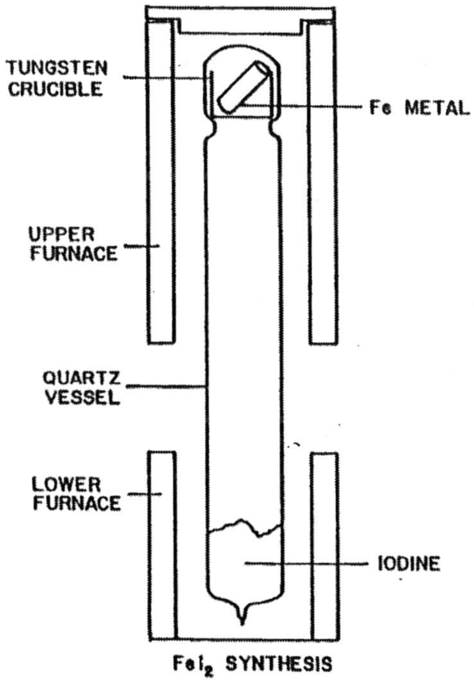

Figure 3. Synthesis of FeI_2 from the elements

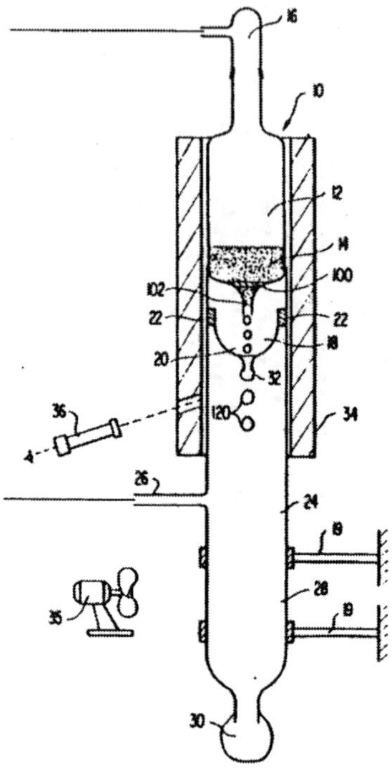

Figure 4. Apparatus for the production of large particles using the "dropping" method

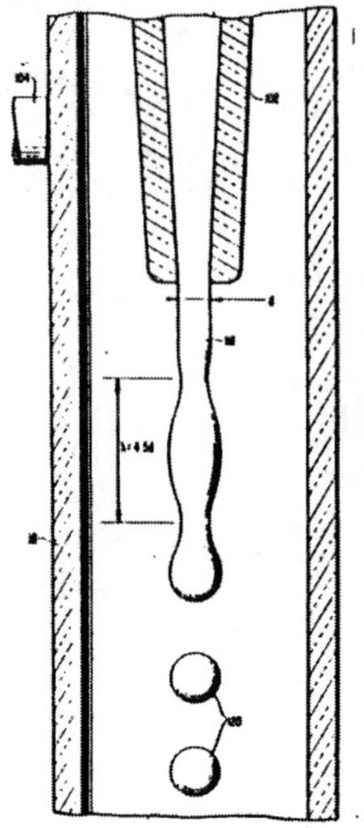

Figure 5. Detail of nozzle for small particle using the "jetting" method

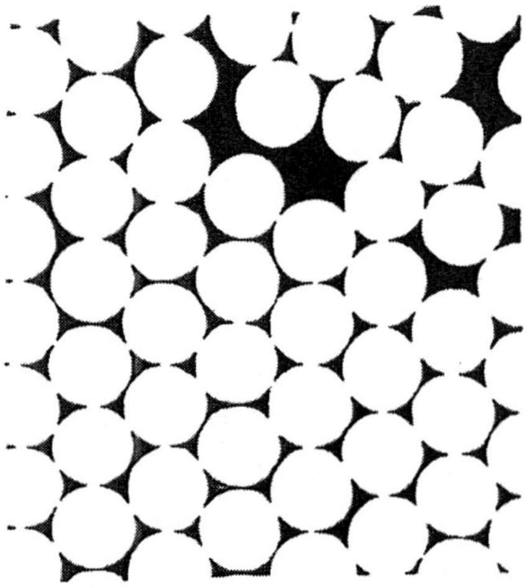

Figure 6. Shadow projection of typical NaI-ScI_3 spheres of diameter 1.8 mm and mass 10 mg.

TABLE I

Spark Source Mass Spectrometric Analysis of NaI, ppm by weight

Li	<0.01	Ca	0.6
B	0.01	Cr	0.04
F	0.5	Mn	<0.02
Na	Major	Fe	0.8
Mg	<0.7	Ni	0.08
Al	0.4	Cu	0.05
Si	0.8	Zn	<0.08
P	0.01	As	<0.02
S	<0.4	Br	0.6
Cl	3	I	Major
K	1		

All other elements except H,C,N and O less than 0.02 ppma (µmol/mol of sample)

ELECTRONICS FOR LAMPS.

Herman Smulders

Advance Transformer Co.

Chicago

Ever since the introduction of the gas discharge lamp, more than 50 years ago, the ballast world has been a magnetic world. The basic functions of ignition, stabilization of the lamp current, and re-ignition of the lamp every half cycle were performed by chokes, transformers and capacitors in a variety of configurations. Over the years the designs changed and the manufacturing methods improved dramatically, but the basics stayed the same: copper or aluminum windings on laminated steel, resulting in a heavy and bulky product, not very efficient, but reliable and the cheapest way to perform those basic functions. This is basically still the situation today, but a change is just around the corner.

More than a decade ago two important facts opened the door for a new way of thinking in the ballast world. The first world energy crisis in the early seventies fixed attention on all possible opportunities of saving energy. And what is more obvious than saving energy in light applications ? So lighting systems, including both lamp and ballast, came under quite some pressure to improve their efficiency. Around the same time semiconductors, capable of handling relatively high voltages and currents, became more and more available as standard products. These two factors started the evolution from the magnetic ballast of the past to the electronic power supply of today.

Why are electronics so attractive for lighting applications ? First of all, electronic ballasts for medium power applications are much more efficient than magnetic ballasts . A magnetic ballast for a 100 watt system normally is

be as high as that of the magnetic ballast.

Today, we are on the verge of the introduction of HI TECH into the traditionally magnetic ballast world; a development which will open the way for more sophisticated ballasts with improved efficiency and which will make possible the introduction of new light sources in new applications.

ENERGY EFFICIENT LIGHT SOURCES

R. Schäfer
Philips Lighting Division, 5600 MD Eindhoven, The Netherlands

I. Introduction

Energy saving has always played an important role in lamp development. Already the basis of the lighting industry, the incandescent lamp, has a history marked by efficacy-milestones. During the first 60 years its efficacy increased from 1.4 lm/W of the first carbon filament lamps to about 14 lm/W of Krypton filled coiled-coil tungsten filament lamps. More recently two further advances have been made, firstly the halogen incandescent lamp introduced around 1958 and secondly the impact on the efficacy studied during the last few years, the IR-reflecting coatings. The latter approach reduces the largest loss mechanism in incandescent lamps: radiated infrared, which is typically more than 70% of lamp input power. With this method efficacy gains of more than 30% have been reported yielding up to 36 lm/Watt at 2000 hr lifetime.

Another way to demonstrate the continuous efforts of the Lighting Industry is to consider the growth of average luminous efficacy of lighting systems. This value can be obtained by dividing the total yearly light production by the respective energy consumption. Before the oil crisis this figure amounted to about 45 lm/W whereas the present value is 54 lm/W.This increase of about 20% has been realized by efficacy improvements of existing lamps and ballast types as well as the replacement of lamps by more efficient ones. An important point in this context is the choice of the most suitable lighting system for a given application.

The physical basis for the understanding of the improvements of lamp efficacy is the energy balance of the lamp. In the following we will discuss energy balances for a number of various discharge lamp types. Special attention will be paid to the physical interpretations of the most important loss mechanisms.

II. General Aspects of Power Balances of Discharge Light Sources

The most common approach to set up the power balance of a discharge tube with outer jacket is shown in the figure. The electrical input power to the lamp is split-up into electrode losses and power supplied to the discharge column, which is converted into non-radiative losses and discharge radiation. The latter is then divided into UV- (250-380 nm), IR- (740-2500 nm) and visible (380-740 nm) radiation. Whilst IR- and visible radiation leave the outer bulb of the lamp, the electrode- and non-radiative losses as well as most of the UV-radiation (depending on the bulb material) lead to heating of the outer bulb. From now on values for the different parts of the power balance will always be given in a percentage of the input power. Apart from this purely physical power balance, the adaptation of the visible radiation spectrum to the sensitivity of the human eye is important for the luminous efficacy. Therefore colour aspects also have to be taken into account when comparing the efficacies of different lamps.

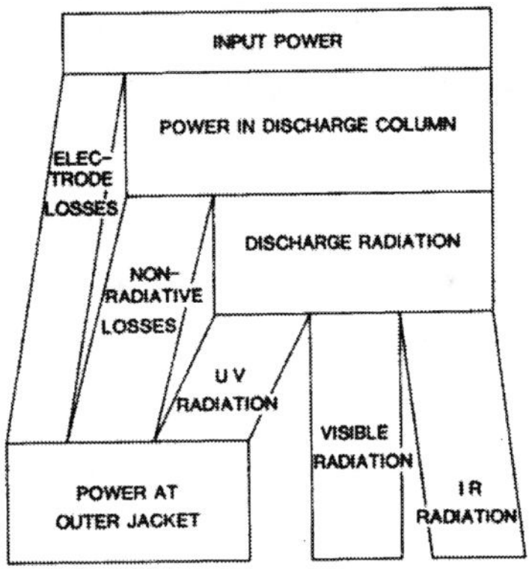

III. Low Pressure Lamps

Among the low pressure lamps we can distinguish two important types, the fluorescent lamps where UV-radiation is converted into "white" visible radiation by phosphors and the low pressure sodium lamp where excited sodium atoms directly emit yellow, nearly monochromatic radiation.

Gap in pagination due to formatting issues.

Pages 53-54

operation. This is due to both a decrease in electrode losses and an increase in column efficiency.

IV. High Pressure Lamps

Among the high pressure discharge lamps the most efficient ones are the high pressure sodium lamps with an arc tube of Na-resistant pca material and the metal halide lamps in a quartz vessel.

1. High Pressure Sodium Lamps

For 400W high pressure sodium lamps the energy balance shows that about 50% of lamp power is converted into discharge radiation whereas electrode losses (6%) and non-radiative losses (44%) account for the rest. The lower non-radiative losses compared to the LPS-lamp are sufficient to keep the much smaller discharge vessel at its operating temperature of 1100-1500 K. The discharge radiation is distributed into UV (0.5%), IR (29%) and visible radiation (29,5%). For optimizing the balance between visible radiation and losses (radiation trapping, heat conduction and IR-radiation) the most important parameters are discharge radius, sodium and buffer gas pressures, wall temperature and power input per unit length. These parameters are not independent of each other so that, in practical lamps, one is not entirely free as to the choice of parameters.

Another important point when determining the efficacy of HPS-lamps is the spectral distribution of the emitted light. The highest efficacy of more than 120 lm/W can be reached with light having a colour temperature of about 2000 K corresponding to a separation of the Na-D line self reversal peaks of about 100 Å. During the last years there has been a trend to go to lamps with higher colour temperatures by increasing the sodium pressure and thus the separation of the self reversal peaks. A consequence of the related better colour quality is a decreasing adaptation of the spectrum to the maximum of the eye sensitivity yielding a lower efficacy.

A further enhancement of lamp efficacy of standard sodium lamps can be expected from heat reflection shields which allow a more independent choice of tube radius, walltemperature and power input. Similarly, electronic power supplies will give more freedom in choosing lamp voltages and currents and thus will reduce the constraints on the choice of optimal discharge conditions. Last but not least new Na-resistant tube materials could lead to more efficient lamps.

2. Metal Halide Lamps

The power balance as well as the luminous efficacy of metal halide lamps depend strongly on the type of emitting species in the arc. Therefore, for 400W lamps the non-radiative losses vary between 27 and 39%, UV-radiation between 3 and 12%, visible radiation between 23 and 34% and IR-radiation between 9 and 27% when comparing e.g. Dysprosium- , Scandium-, Na-Tl-In- and Tin-halide discharges. This is mainly due to the different absorption and emission mechanisms of the constituents. On the whole, rare earth metals have turned out to be the most efficient additives yielding luminous efficacies of more than 110 lm/W for white radiation. Further improvements of metal halide lamps can be expected from reduction of axial segregation e.g. by a suitable choice of buffer gas pressure and arc radius.

V. Comparison of the Different Lamp Types

The most relevant data of the four lamp types discussed in this paper are presented in the following table.

lamp type	electr. losses	non-rad. losses	radiation UV	VIS	IR	TOTAL	lm/W
LPM 36W	16%	18%	0.5%	28%	-	63%	96
LPS 180W	12%	50%	-	35%	3%	38%	180
HPS 400W	6%	44%	0.5%	29.5%	20%	50%	120
MH 400W	10%	27-39%	3-12%	23-34%	9-27%	52-64%	60-110

We see that the somewhat paradoxical situation arises that the LPS-lamp with the highest luminous efficacy and visible radiation efficiency shows the biggest amount of non-radiative and electrode losses. The reason for this behaviour is the very good matching of the nearly monochromatic sodium radiation to the maximum of the eye sensivity as well as the low generation of IR-radiation. Most important for future efficacy gains is therefore a further reduction of electrode- and non-radiative losses, e.g. by better heat insulation of the discharge tube.

A quite different behaviour can be found for fluorescent lamps. Here electrode and non-radiative losses are lowest and discharge radiation efficiency is highest. Nevertheless the luminous efficacy is low due to only about 40% energy conversion efficiency of present one-photon-phosphors. Therefore the biggest improvements in this lamp type might be reached by developing two-photon-phosphors with a high quantum efficiency.

Another important loss mechanism for the high-pressure lamp types is IR-radiation, in addition to electrode- and non-radiative losses. A further efficacy increase requires reduction of IR- an non-radiative losses. This might be achieved, as already mentioned, by reducing constraints on discharge optimization owing to the application of electronic power supplies, IR-reflecting layers or electrodeless discharges.

MODELING OF LOW POWER HP DISCHARGE LAMPS

E. Fischer
Philips GmbH Forschungslaboratorium Aachen
D-5100 Aachen, Federal Republic of Germany

1. Introduction

The development of theoretical lamp models may have two different aims. One is to obtain a quantitative understandig of the processes that determine the properties of the light source, the other is to give ingeneering aids for the construction of new or improvement of existing lamp types. For the latter case the model should describe as accurately as possible the complete system formed by the lamp plasma, the envelope and the ballast, possibly including also the outer bulb and the luminaire, as all these parts interact thermally or electrically. In physical models it may be of advantage to use idealized and simplified conditions to study individual phenomena more clearly and without superfluous effort.

Thus all the numerical lamp models developed by now only consider one aspect of the lamp. Either the properties of the plasma are investigated, mostly idealizing the arc as a part of an infinitely long cylinder, or the processes at the electrode surface are studied regarding the plasma state in the arc column as a boundary condition at infinity.

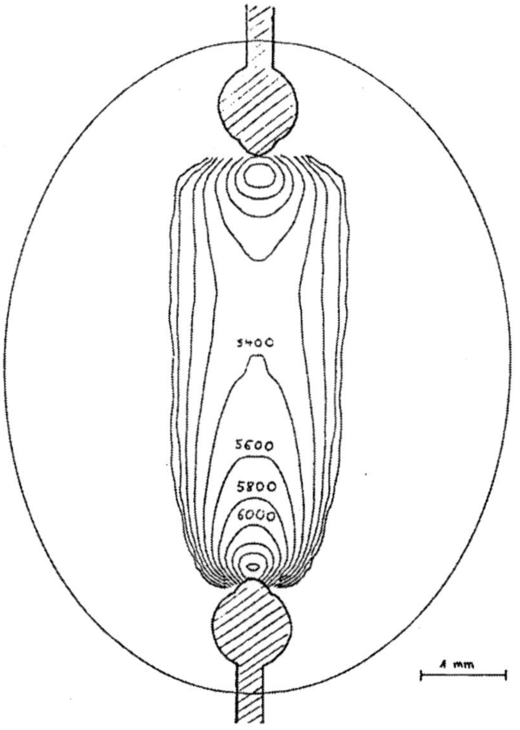

Fig. 1 Measured field of isothermes for a Hg arc lamp (P_{Hg}=30 bar, L_{el}=35 W)

In low power HP discharge lamps, however, the coupling between the eletrode regions and the arc centre is so strong that it is no longer sensible to use such idealized models. This can easily be understood looking at the plasma temperature profile of such a low power lamp as it is shown in fig.1. The figure shows the temperature field in a 30 W mercury lamp burning in an elliptical envelope. It clearly shows up that the temperature is not constant along the arc axis but rises towards the electrodes reaching its maximum immediately in front of the electrode tips. As a consequence the electrode regions contribute essentially to the radiation output of the lamp. Also the electrical data, i.e. the arc voltage and the reignition peak (for a.c. operation) are determined by the properties of the electrode regions.

To understand the phenomena occuring in such lamps the aim of our work therefore must be to set up a consistent numerical model that allows the determination of the local plasma properties from the arc centre to the electrode surface and the thermal data of the electrode itself.

2. Properties of HP Discharge Lamps

Before we come to the model calculations let us first have a look at the basic properties of high pressure discharge lamps. Common to all these lamp types is that the main constituent of the plasma is mercury vapour, in low power lamps ($\leq$ 70 W) typically at an operation pressure of 10-30 bar. Various other chemical species may be present to improve the lamp properties, but due to the high mercury concentration all transport processes in the plasma are dominated by collisions of some other particles with neutral mercury atoms and the mean free path of the electrons is so short that it is comparable to the mean distance of the atoms. As a consequence the energy take up of electrons from the electric field between two collisions is so small that all secondary emission or ionization processes that play a role in low pressure discharges are unimportant in a high pressure lamp plasma even at the highest field strengths that may occur in the electrode boundary layers.

Thus in spite of the very complex chemistry that has to be considered especially in metal halide lamps, the HP discharge lamp is simple from the plasma-physical point of view. At least as a first approximation the assumption of local thermal equilibrium (LTE) all over the arc volume should lead to an acceptable description of the plasma properties.

3. Simplified Arc Model

As a first step we will consider a simplified LTE model. In addition d.c. operation is assumed, but all segregation effects that may be caused by electrical or buoyant forces are neglected. Convective energy transport is not taken into account, too. Under these conditions the energy balance of the plasma reads as follows

$$\mathrm{div\,grad}\, S + \sigma (\mathrm{grad}\, \Phi)^2 - U_R = 0 \tag{1}$$

where S is the heat flux potential defined as

$$S = \int_{T_W}^{T} \lambda \cdot dT \ , \tag{2}$$

T_W being the wall temperature (assumed as constant) and λ the thermal conductivity depending only on the local temperature T. The same holds for the electrical conductivity σ. Φ is the electric potential and U_R the net local emission of radiation. The most important radiative energy transport is in the form of optically thick UV mercury lines. As the mean free path of these photons is very short, this energy transfer acts like an additional thermal conductivity. Thus the radiation term can be ommitted if instead a radiative contribution is added to the thermal conductivity. Details on the calculation of the transport coefficients $\sigma(T)$ and $\lambda(T)$ from the chemical equilibrium and radiation transport data can be found elsewhere /1-4/.

The energy balance (eq.1) has to be solved simultaneously with the electrical field equation

$$\mathrm{div}\, \vec{j} = -\mathrm{div}(\sigma \cdot \mathrm{grad}\, \Phi) = 0 \ . \tag{3}$$

The boundary conditions for the electric potential result from the fact that the electric current through the lamp wall must be zero and that the potential has a constant value on the electrode surface. The heat flux potential S must be zero at the wall, its value on the electrode surface results from the energy balance of the electrode. We will come back in detail to this point later. As a first approximation we can confine ourselves to the most important feature of the electrodes - besides that of being the source at the electric current - namely to be a very efficient heat sink, as their temperature is low - low at least compared to the temperatures in the plasma. Thus the condition $S = 0$ on the electrode surface will be a good approximation.

Numerical solutions can be obtained by finite element methods. For special geometries more accurate and less time consuming semi-analytical procedures may be applied. The numerical results given here have been obtained for elliptical lamps. For this case it was convenient to use series expansions of the variables with respect to the eigenfunctions of the Laplace operator in elliptical coordinates (Legendre functions of fractional order) and replace the real electrode surface by a hyperboloid of same curvature at the tip. The advantage of this method is that the determination of the eigenfunctions which depend only on the lamp geometry must be performed only once. The iterative determination of the temperature field - iteration is necessary due to the highly non-linear dependency between the electrical conductivity and the heat flux potential S - can then be done simply be solving ordinary integrals. Details of the numerical method are described in /1/.

Fig.2 shows a field of isothermes as calculated from this simplified model. It shows the typical arc constriction and hot spot formation that is observed in the experiments (see fig.1).

Thus we can conclude that it is predominantly the cooling of the plasma by the electrodes that causes the arc to form a hot plasma spot near the surface to compensate for the additional energy loss. Variation of the electrode geometry shows that enhancement of the field strength in front of a narrow electrode tip is favorable for the formation of small spots, but constriction and spot formation also occur with nearly flat electrodes.

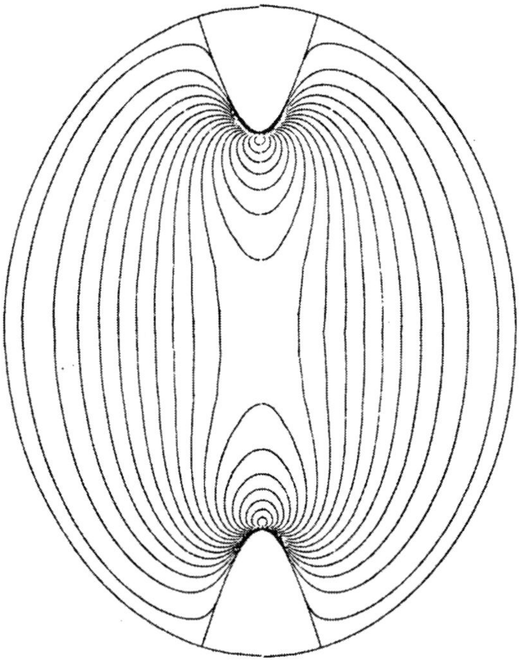

Fig. 2 Calculated field of isothermes (LTE model)

The calculated data yield not only the temperature distribution, but also the arc voltage and the heat flux to the wall and to the electrodes. But while the plasma temperature and the heat flux to the walls agree well with experimental data, the heat flux to the electrodes and still more the arc voltage show very large discrepancies. This is caused to some extent by the assumption of S = 0 on the electrode surface. This assumption means that the electrical conductivity is practically zero at the surface. Thus the electric field had to be infinite to carry the arc current through this non-conductive layer. Assuming a more realistic temperature on the electrode surface reduces the discrepancy slightly, but does not bring the results close to the experimental data. This can be reached only by means of a more detailed model that takes into account non-equilibrium effects at the electrode boundaries.

4. Non-equilibrium Model

The main reason for the failure of the equilibrium model presented in the last section is that the electrical conductivity values in the boundary layer at the electrodes are too low. The results show, however, that the temperature gradients in this boundary layer are extremely high (up to 10^9 K/m in the example of fig.2). Thus we must expect strong diffusion of charged particles from the plasma to the electrodes, which leads to an enhancement of the electrical conductivity in this zone. To determine the electron density in the boundary layer we must solve the diffusion equations for the electrons and ions under the influence of the electrical forces caused by the particles themselves and by the outer electric field. For a weakly ionized plasma in which diffusion is dominated by collisions with neutrals (in our case Hg atoms), the diffusion equations have the form /5/

$$\frac{1}{n^2\mathcal{D}_{eo}} (n_e \cdot \vec{j}_o - n_o \cdot \vec{j}_e) = \text{grad}\,\frac{n_e}{n} + \frac{e n_e \vec{E}}{p}$$
$$\frac{1}{n^2\mathcal{D}_{io}} (n_i \cdot \vec{j}_o - n_o \cdot \vec{j}_i) = \text{grad}\,\frac{n_i}{n} - \frac{e n_i \vec{E}}{p} \qquad (4)$$

These equations relate the particle densities to the corresponding fluxes, which are connected to each other by the conservation laws for mass and electric charge

$$\vec{j}_o = -\vec{j}_i \qquad (5)$$

and

$$\text{div}\,\vec{J} = e \cdot \text{div}\,(\vec{j}_i - \vec{j}_e) = 0, \qquad (6)$$

where $\vec{J}$ is the net electric current density.

In the continuity equations for the individual particle species it is necessary of course to consider production and annihilation. Production of ions and electrons takes place predominantly by collisional ionization with a production rate given by

$$R^+ = q^I \cdot n_e \cdot n_o \quad . \qquad (7)$$

Loss of charged particles is mainly by radiative recombination and by three-body recombination with a neutral atom being the third particle:

$$R^- = q^R \cdot n_e n_i + q^S \cdot n_e n_i n_o \quad . \qquad (8)$$

q^I, q^R and q^S are the rate coefficients corresponding to the three processes. For practical use we can make use of the fact, that in LTE plasma production and annihilation must cancel. So introducing the equilibrium density of charged particles n_{io}, the rate equation can be written as:

$$\text{div}\,\vec{j}_i = q^I \cdot n_e \cdot n \cdot \left(1 - \frac{n_i}{n_{io}}\right) \quad . \qquad (9)$$

Combining this equation with the diffusion and continuity equations and making use of the fact that in a weakly ionized mercury plasma the electrical conductivity is proportional to n_e/n, we find an additional differential equation that relates the actual conductivity field to its equilibrium data:

$$\text{div grad}\,\sigma = \frac{q^{I} \cdot n}{2\mathcal{D}_{io}} \cdot \frac{\sigma}{\alpha}\left(1-\alpha\frac{\sigma}{\sigma_0}\right) , \qquad (10)$$

where σ_0 is the equilibrium value of the conductivity as used in the simplified model, and α is the ratio n_i/n_e. The space charge distribution can be calculated from the electical field equations

$$\text{div}\,\varepsilon\,\text{grad}\,\Phi = e(n_e-n_i) \quad \text{and} \quad \text{div}\,\sigma\,\text{grad}\,\Phi = 0 \,. \qquad (11)$$

The set of equations given above is valid both for the anode and cathode region. The only difference comes in from the boundary conditions at the surface. While at the anode surface the ion current must be zero, there may be both ion and electron current at the cathode, but here the electron current density is limited by the maximum emissivity of the surface. Fortunately here again we can take advantage of the special properties of the high pressure mercury plasma. Due to the high collision rate between charge carriers and neutrals, secondary emission processes from high energy particles originating in the plasma are negligible, as even at field strength values that would cause field emission from the surface, the energy take up of electrons from the electric field between two collisions is by far less than the work function of the electrode material. So we can use the well known emissivity formulas given in literature (see e.g. /6/), but not only thermionic emission must be taken into account but also enhancements due to Schottky effect, thermal field emission or possibly even pure field emission.

The determination of the emission current requires exact knowledge of the electric field strength at the surface and of the surface temperature. While the electric field results from the solution of the basic set of equations discussed above, the surface temperature must be determined in a separate step by solving the energy balance of the electrode. In practical lamps various types of electrodes are used: rods, hollow coils, coils with a central rod or rods with spheres fused into the top. For the model calculations we have chosen the form shown in fig.3. The model electrode consists of a cylindrical rod with a sphere on top and a second half sphere of different radius on the top of the first one. This construction has the advantage that the relevant parameters of most real electrodes can be simulated by a suitable choice of the radii of the spheres and of the dimensions of the rod.

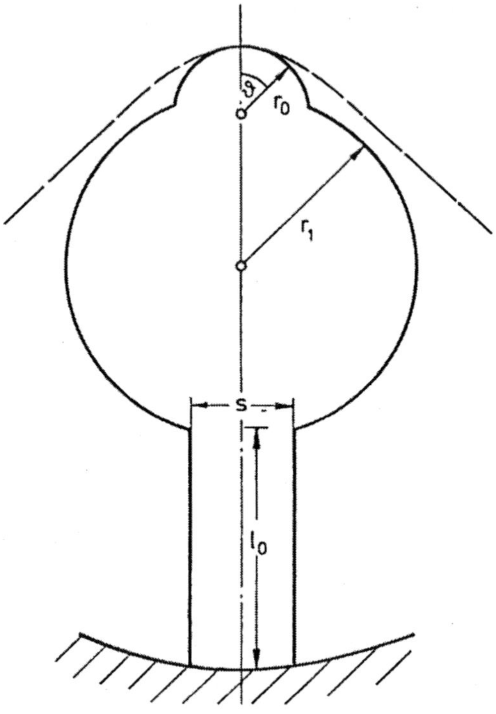

Fig. 3 Shape of a model electrode

The radius of the upper sphere determines the field distribution near the surface, the size of the lower sphere determines the radiative loss from the electrode surface and the rod dimensions control the conductive heat loss to the base point. Energy transfer from the plasma is concentrated to the vicinity of the plasma spot, for the model electrode to the surface of the upper half sphere. The heating and cooling mechanisms that must be considered are heat conduction and radiation from the plasma, heating of the anode and cooling of the cathode by the electron current, heating of the cathode by the ion current and the radiative loss of energy from the electrode surface:

$$q(\vartheta) = -(\text{grad } S)_n - \varepsilon\sigma T^4 + (E_{eff} \cdot j_e + (E_i - E_{eff}) \cdot j_i) \qquad (12)$$

where j_e has to be taken negative at the cathode, E_{eff} is the effective work function, and E_i is the ionization energy (or its weighted mean value in a gas mixture). Radiation is the only relevant energy loss mechanism on the surface of the lower sphere and of the rod.

The model outlined above gives a consistent description of high pressure discharge lamps including the electrode regions of the plasma and the electrodes themselves. Due to the complexity and the non--linear connections between the variables it is best to solve the basic system by an iteration method similar to that mentioned for the simplified model. A few additional steps must be included, however, in the procedure. A schematic representation of the way of solution is given in fig.4.

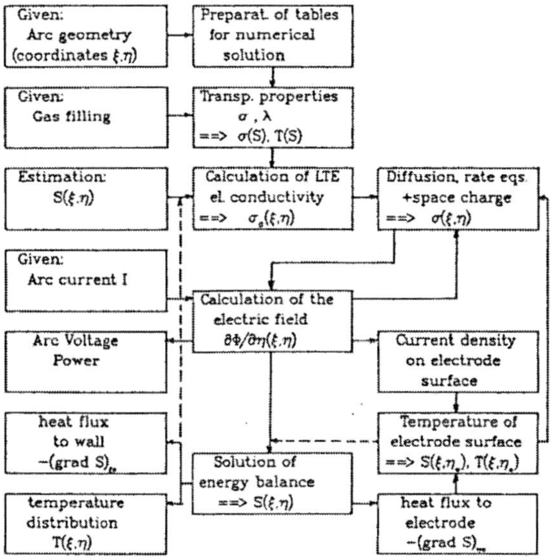

Fig. 4 Scheme of numerical solution for the non-LTE model ξ and η are coordinates perpendicular to the lamp wall resp. to the electrode surface.

5. Results

The arc model derived in the foregoing chapter allows the description of high pressure discharges, where the plasma is collision-dominated in the electrode boundary layer as well. For such arcs it delivers not only the temperature distribution in the plasma, but also the voltage-current characteristics, the energy flux to the wall, the energy loss to the electrodes and the temperature and electric field distribution on the electrode surfaces. The model contains no adjustable constants, but is based solely on fundamental data of the gas contained in the arc. Thus the results can only be as good as the input data, which are not all known with high accuracy. An exact coincidence of the results with measured data cannot be expected, but on the other hand the relatively good agreement in the data is a strong indication that at least the most relevant physical effects have been taken into account in the right way.

As an example of the good agreement between model calculations and measured data fig.5 shows some current-voltage-characteristics of mercury arcs. The various curves in the graph correspond to different curvatures of the electrode tips. The experimental curve belongs to a lamp with electrodes etched from a tungsten bar to a tip diameter of about 350 µm. The operating pressure of 30 ± 3 bar was not measured but estimated from the lamp filling and from the calculated temperature profile.

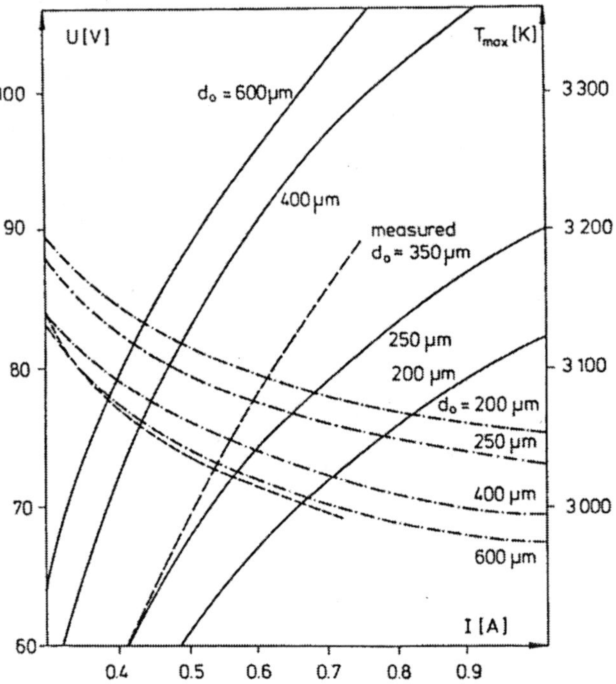

Fig. 5 Voltage-current characteristics (—·—·—) and cathode temperatures (———) of mercury arcs (p=30 bar) with various tip diameters. Experimental data are shown as dashed lines.

The second set of curves in fig.5 gives the tip temperature of the cathode calculated for a model electrode according to fig.3. The results show that although the arc voltage - and also the cathode fall - are higher for smaller tip diameters, the heat loss to the electrodes and thus also the tip temperature are lower. This result shows that conduction and radiation from the plasma are the most relevant heating mechanisms in the high-pressure mercury arc, whereas ion bombardment, which predominates in low pressure discharges, is less important. The measured curve shown in the graph also belongs to the lamp mentioned above. The agreement with the numerical results is quite satisfactory having regard to the fact that for the absolute value of the electrode tip temperature an error margin of at least ± 200 K must be admitted to allow for disturbances of the pyrometric measurements by stray light from the plasma and for uncertainty of the surface emissivity.

The discussion in the last section serves only to illustrate the reliability of the data obtained from the model. Results for special arc parameters will not be discussed in more detail here. Instead another form of presentation of numerical results will be given below that allows somewhat more insight into the phenomena occurring upon variation of the arc parameters. For that purpose the scheme of solution has been changed slightly. Instead of varying all the quantities in the iteration process, the field strength at the electrode surface has been regarded as an independent parameter in the determination of the electron current and in the solution of the diffusion equations. The field strength value at the surface resulting from the field equation can then be plotted as a function of the input value. Such a graph is shown in fig.6. The true solution is then given by the point where the curve crosses the diagonal. Fig.6 shows some curves for a variation of the cathode tip diameter. All the curves exhibit a wide range in which the slope is very close to one. Varying the tip diameter leads to a shift of this nearly parallel part from above to below the diagonal. Thus the position of the crossing point may change substantially with small variations of the tip diameter. The field strength at the electrode changes by three orders of magnitude in this transition from values of the order of up to 10^9 V/m, which is close to the pure field emission regime, to field strengths of some 10^5 V/m,

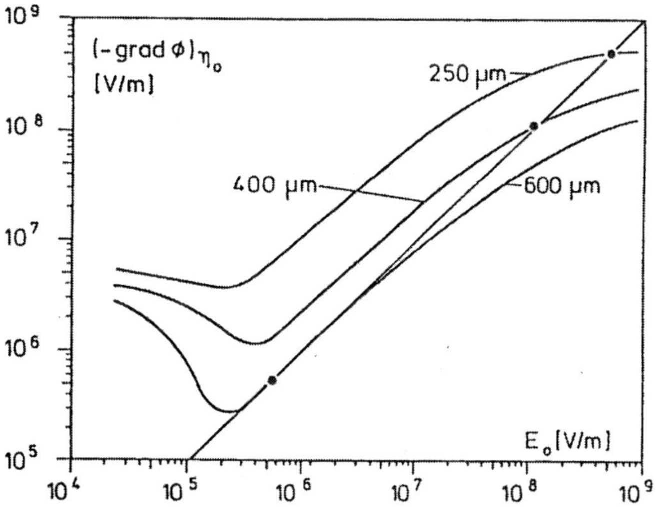

Fig. 6 Determination of the electrical field strength at the cathode surface dependent on electrode tip diameter (Hg, 30 bar, I = 0.45 A)

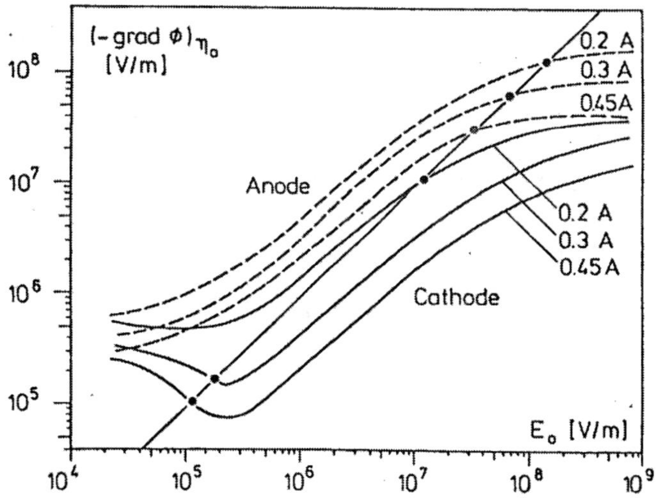

Fig. 7 Determination of the electrical field strength at the electrode surface dependent on arc current for cathode (——) and anode (– – –) of mercury arcs (p = 30 bar)

where only thermionic emission and the Schottky effect must be considered.In this burning mode the ion current makes up a reasonable fraction of the total current, while in the high-field mode the electron current is dominant.

Similar abrupt changes in the burning conditions also occur upon variation of other arc parameters as the heat resistance of the electrode or the arc current. In fig.7 curves for various arc currents are plotted. Here again the crossing point changes from the high-field mode to the low-field mode with increasing current. Fig.7 also contains curves for the anode. The curves have a somewhat similar shape except that they do not exhibit the sharp kink in the low-field region that occurs in the cathode curves due to the influence of the ion current. Though in principle strong variations in field strength may also be expected at the anode, in the parameter range covered by our experimental lamps the solutions were always of the high-field mode.

A qualitative demonstration of a transition between the two burning modes of a cathode spot is given in fig.8, which shows the voltage of an a.c. operated lamp with a sinusoidal arc current. If the temperature of the cathode is high enough, the arc voltage will show a smooth behaviour as shown in the upper photograph. The arc burns in the low-field mode all the time. If the arc power is decreased, the tip temperature is lowered. After current reversal the arc first starts in the high-field mode. With increasing current the conditions on the electrode change. Higher current and higher tip temperature favour the low-field mode. Within some tenths of a millisecond the arc switches to this mode. The transition shows up in the time dependence of the arc

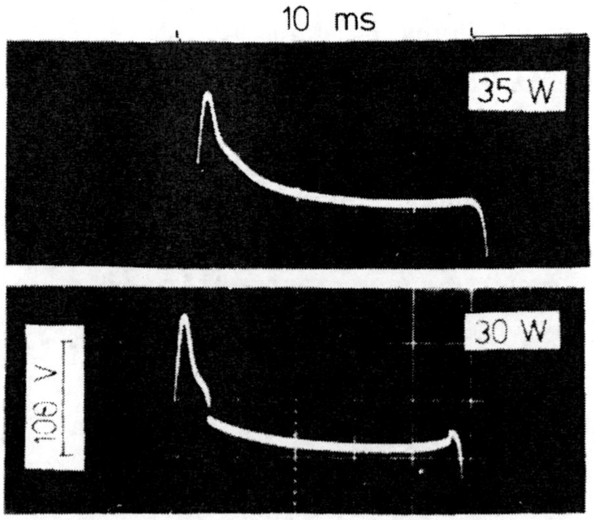

Fig. 8 Arc voltage vs. time for a.c. operated lamps (50 Hz).

voltage as a sudden drop of the order of 10 V (fig.8, bottom). The model presented in this report can of course only give a qualitative explanation of the transition phenomena; a quantitative description of the effect must await an extension of the model to a.c. operated lamps.

The examples discussed up to now were all concerned with the influence of geometric or thermal conditions at the electrodes on the properties of the discharge. The fact that the composition of the arc plasma also influences the shape of the hot plasma spot in front of the electrodes is demonstrated in fig.9. The upper part shows the field of isotherms in the cathode region of a mercury arc lamp, the lower part shows an arc in otherwise identical conditions, but with the addition of NaI, TlI and InI. In the latter case the cathode spot is much less constricted. This difference, which also results in a lower temperature of the electrode tip, is observed in experiments, too. That the reduction of the tip temperature, which should be of the order of 300 K according to the model calculations both for the cathode and for the anode, is much smaller in real a.c. lamps, can be explained by the fact that during the anode half cycle most of the sodium that is nearly totally ionized is removed from the anode boundary region by diffusion in the electric field, so that the arc is more like a mercury discharge in this half cycle.

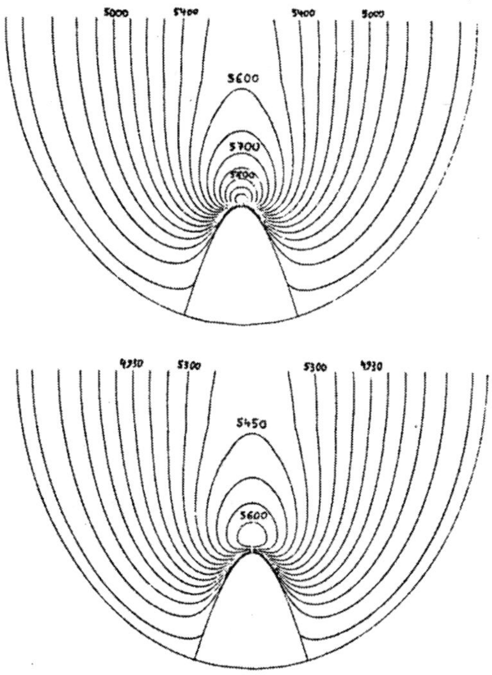

Fig. 9 Comparison of temperature profiles for a pure mercury arc (top) and an arc with addition of NaI, TlI and InI (bottom). (p = 30 bar, I = 0.45 A)

For the practical design of discharge lamps a question of importance is what the heat resistance of an electrode must be to achieve a given tip temperature and what burning voltage and energy losses to the electrodes will occur under these conditions. Fig.10 shows a graph of the arc voltage versus the tip temperature for the cathode of a mercury arc. As can be read from the curve, any attempt to reduce the tip temperature will result in an increase of the arc voltage. The additional power introduced into the plasma will be conducted to the electrodes. The dependence of the energy loss to the cathode on the tip temperature is shown as a dashed-dotted line in the graph. The plasma counteracts an enhanced cooling by an increase in the amount of energy that is delivered to the electrode surface from the plasma. Cooling to below a certain limit, which is about 2660 K in the example given here (E_A = 4.5 eV), is practically impossible. The attempt to remove more heat from the cathode would only have the effect of increasing the energy transfer from the plasma by the same amount.

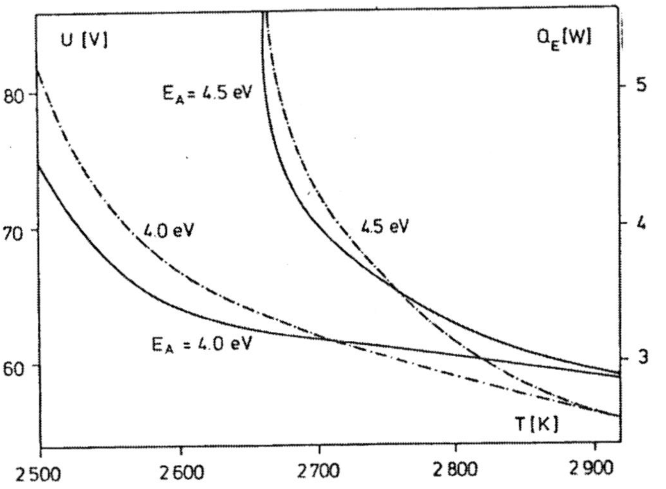

Fig. 10 Variation of arc voltage (———) and electrode loss (—·—·—) with electrode tip temperature (Hg, p=30 bar, I=0.45 A) for two values of the cathode work function

A further reduction of the tip temperature below the limit is possible only by reducing the work function of the electrode material, as demonstrated by the second set of curves shown in fig.10. A reduction from 4.5 eV to 4 eV would lead to a lowering of the limiting temperature by 200 K. The results shown in fig.10 can also be looked at from another point of view, however. The model calculations show that in small high-pressure discharge lamps (30 W in the example shown here) the electrode losses are about 20%, even if the tip temperature is very high. Thus a further increase of these losses would considerably reduce the efficacy of the lamp, and therefore one should not try too hard to reduce the tip temperature below 2800 K, as it would make the lamp less efficient. If however a tip temperature of 2800 K is tolerable, it does not make much sense to look for emitters to reduce the work function, as in this temperature range a change of the work function has only a minor influence on the energy loss of the electrode.

Though the model presented here is far from being complete in the sense of an ingeneering model, many new insights relevant to the construction of low power HP discharge lamps can be deduced from the calculations, and the results may serve as a base for more detailed studies of chemical transport processes or of the radiative properties of such lamps.

7. References

1. E. Fischer, Proc. XVII Int. Conf. on Phen. in Ionized Gases, Part II (Inv. Lectures), Budapest (1985)
2. E. Schnedler, CALPHAD, Vol. 8, 265 (1984)
3. H.P. Stormberg, J. Appl. Phys., 51, 1963 (1980)
4. R.J. Zollweg, R.W. Liebermann and D.K. McLain, J. Appl. Phys. 52, 3293 (1981)
5. J.O. Hirschfelder, C.F. Curtiss and R.B. Bird, Molecular Theory of Gases and Liquids, Wiley, New York (1964), pp. 514-520
6. W.P. Dyke and W.W. Dolan, in Adv. in Electronics and Electron Physics, Vol. VIII, Acad. Press (1956)

Nr. H

COMPUTER AIDED CHEMISTRY FOR LAMP DESIGN

E. Schnedler
Philips GmbH Forschungslaboratorium Aachen
Postfach 1980, D-5100 Aachen, Federal Republic of Germany

ABSTRACT

Several computer models of high temperature chemistry have been used to describe transport phenomena in lamps. These models are reviewed and their applicability as design tool is discussed. The calculation of a complex chemical equilibrium is utilized as an example for computer aided lamp chemistry. More realistic models consider transport effects. Their application to halogen incandescent lamps is demonstrated. Effects which still have to be considered along with computer aided chemistry for lamp design are discussed.

I. INTRODUCTION

High temperature lamp chemistry is a very complex matter. The filling of lamps undergoes reactions with impurities like water or oxygen, with bulb materials like silica or alumina, and with the electrode or filament material tungsten or possible additives enhancing emission. The number of known compounds which can be formed by the elements contained in the filling, in impurities, in electrode and bulb materials, often amounts to several hundreds. Already from the high quantity of possible compounds it is obvious, that it is extremely difficult to predict relevant reactions, which may take place in an operating lamp.

Considering physical transport effects together with chemical reactions results in an even worse situation. Due to concentration- , thermal-, and ambipolar diffusion as well as convection, material can be transported through the lamp vessel. Heavy molecules, like WO2I2, with small diffusion coefficients can be formed in cooler regions near the bulb, whereas lighter molecules, atoms, and ions, like W or WO, with high diffusion coefficients are formed in hotter parts of the lamp. This leads to segregation of chemically reactive gases, which in turn influences the possible reactions.

But even the calculation of a "simple chemical equilibrium with only a few compounds" seems to be too tedious to be accomplished without the help of a computer; beyond that, realistic modeling of chemical transports in light sources, including convection, concentration-, thermal-, and ambipolar diffusion, reaches the limits of capacity of modern powerful computers.

During the last twenty years software has been developed to understand the complexity of transport reactions. Today it seems possible that modeling of high temperature chemistry may become a tool for the design of lamps.

II. PAST AND PRESENT MODELING TECHNIQUES

II.a Calculation of chemical equilibria

The common procedure of nearly all applications of equilibrium calculations to the description of transport phenomena in halogen incandescent lamps is to assume that solid tungsten is present at all temperatures from the bulb temperature to the maximum filament temperature. The chemical equilibrium resulting from all possible reactions of tungsten with the halide additives of the filling and with impurities like oxygen or water is calculated at all considered temperatures. The vapour pressure of tungsten and the partial pressures of tungsten-containing compounds are summed up to the so-called tungsten solubility SpW, or, if given as function of the temperature, the so-called SpW(T) curve. The calculations start from the same pressure values of reactive gases everywhere on the wall and along the filament. This assumption is by no means trivial and we will come back to it in the following discussion.

As tungsten is transported, due to diffusion, from regions of high solubility to regions of lower solubility, the SpW(T) curve can give informations about the transport behaviour arising from the chemical filling.

The desired direction of the tungsten transport cycle in a halogen incandescent lamp naturally is from cold to hot sites, i.e. a filling shall bring back any trace of metallic tungsten from the lamp wall to the incandescent filament, thereby preventing the bulb from blackening. In other words, it should fulfil the requirement SpW(Tb) > Spw(T) for all T > Tb.

This model was introduced by Kopelman and van Wormer to explain transport mechanisms in halogen lamps containing bromine and iodine (1), and in lamps containing chlorine, fluorine, and hydrogen (2). Yannopoulos and Pebler used this method to describe the W-O-C-H-Br sytem (3,4).

In (5) a review of the applications of the model of constant stoichiometry on incandescent lamps has been given, which is as far as I can see by no means complete. The descriptions reviewed suffer from the neglection of condensed phases, which might reduce the reactive filling of the gas phase.

Geszti (6) considered diffusion and convection together with chemical reactions for transports in halogen lamps. But he only was able to solve the resulting equations under the assumption of equal diffusion coefficients for all chemical species, and thus ended up again with a model of constant stoichiometry in the total reaction volume. Additionally in (6) the problem of different condensed phases at sites of differing temperatures is pointed out for lamps containing C, Br, O, and H as filled elements.

Van den Hoek and Rouweler (7) examined the influence of different diffusion coefficients of the molecules involved. They showed that the assumption of constant stoichiometry is justified as long as the binary diffusion coefficients of all atoms and compounds coincide.

Passmore, Duggan, and Waymouth (8) discussed the effect of condensed solid or liquid compounds at the cold bulb wall of a lamp containing impurities (H_2O, O_2, C) and phosphorus additives to the filling gases N_2 and Ar. The active filling is reduced by the amounts of the condensed phases, and thus the tungsten transports are reduced.

Recently a paper by Spacil and Luthra (9) has shown that chemical equilibrium calculations are still a very useful tool to discuss the chemical transport reactions which take place in halogen incandescent lamps.

These calculations are of major importance for the understanding of high pressure discharge lamps, too, and a number of papers dealing with this problem have been published.

Some calculations concerning a small HPD lamp filled with NaI, InI, TlI, As, and Hg are given by Fischer (10). He shows that in these lamps it is possible to establish a regenerative tungsten iodine cycle, which can prevent the bulb from blackening if the amounts of impurities are controlled properly.

Van Erk (11) shows, that it is possible to understand the complex transport reactions of silica and silicon in rare-earth metal halide lamps. Due to reactions of rare-earth iodides with quartz, silicon compounds can occur in the gas phase and, after disproportionation, either be transported as silicon to the tungsten electrode, or be deposited as quartz at a cooler spot of the lamp.

II.b A constant stoichiometry model

As was already pointed out (8) - (11), the reactive gases which have been filled into a lamp, may alter due to condensation of solid or liquid compounds. I.e. even if we assume that a constant stoichiometry exists all over the gas phase, the stoichiometry of this gas phase may have been changed from the filled situation by the condensation of compounds.

In general one has to consider all sites of different temperatures where condensed phases can be present. However very often the formation of condensed compounds is connected with an exothermic reaction. This means, that the coolest site in the lamp is the most probable region for the deposition of condensates. Therefore it appears evident to perform the calculation of chemical equilibria in lamps in two steps:

First calculate the chemical equilibrium with the total filling and expected impurities at the temperature of the coolest site in the lamp. Determine the condensed phases which can exist under these conditions and calculate the composition of the gas phase above these condensed compounds.

Second take this composition as filling for the equlibrium calculations at higher temperatures, from which the solubility curves of tungsten or of the bulb materials can be evaluated.

The following example may demonstrate this procedure. It is a calculation for a 400 Watt high pressure discharge lamp is shown. Let us assume that the coldest spot is near the foot of the lower electrode at a temperature of about 1100 K. The filling at operating temperature is assumed to be 20 bar Hg, 2 bar NaI, 0.1 bar TlI, 0.01 bar InI (if the filling were fully vapourized), an impurity of 0.1 mbar H2O from the hygroscopic salts, and a small amount of WO2 corresponding to an oxygen pressure of 1e-3 bar. The calculation has been performed with 128 gaseous elements, ions, and compounds, with 33 liquid and 78 solid compounds. Reactions with the quartz of the bulb were admitted.

At the bulb temperature of 1100 K, the oxygen impurities arising partly from tungsten oxide and partly from water react with sodium iodide and tungsten to form liquid Na2WO4 and free iodine.

$$2NaI + W + 2O_2 \longleftrightarrow Na_2WO_4 + I_2.$$

This reaction will change the stoichiometry of the gas phase slightly, depending on the oxygen content of the lamp and on the temperature at the coldest spot. This reaction may have an important effect upon the pressure of free iodine. The sodium iodide content of the gas phase turns out to be 1.7e-3 bar, corresponding to its vapour pressure at 1100 K. Reactions with quartz turn out to be of minor importance.

The tungsten solubilty curve is then calculated with the composition of the gas phase above the condensed phases NaI, SiO2, and Na2WO4 . At 3100 K, the temperature of the hot spot of the electrode, the filling pressure of tungsten is reduced to the tungsten content of the gas phase at this temperature. Calculations at higher temperatures then are carried out as homogeneous chemical equlibria which give information of the composition of the plasma with the concentrations of ions up to 5900 K. Results are shown in Figs. 1.

Figs. 1a-1c show the distribution of the partial pressures vs. temperature. Fig. 1d gives the corresponding SpW(T) curve. One finds that tungsten can be carried from the top of the electrode and from regions with temperatures of about 1900 K to the bulb, an effect which has been observed in such lamps after several thousand hours of operation.

It is not difficult to generalize the described model to models with different temperatures for condensed phases.

II.c Transport effects

Computer models with transport effects published so far were transport - limited, i.e. all authors assumed, that chemical reactions are fast enough to maintain local equilibrium within the lamp. This assumption seems to be justified due to the relatively high temperatures which occur during operation. Nevertheless mass exchange between sites of different temperatures can occur and therefore constant stoichiometry can be violated due to diffent transport coefficients of the various species which exist in different regions of lamps.

A model, which showed that segregation effects of chemically reactive gases are of importance for lamp behaviour, was presented by Fischer (12) for mercury-metal-halide arcs. In this paper the energy balance and the concentration distribution of the reactive gases in a rotationally symmetric discharge lamp have been calculated.

In a subsequent paper (13) Fischer considered the axial segregation of the plasma composition, which was observed experimentally by changes in the emission spectrum along the axis. For an infinitely long cylindrical lamp it is possible to determine the macroscopic convection velocity of the mercury gas.

Stormberg (14), (15) published two papers where thermal and ambipolar diffusion for charged particles were included in a radial diffusion-convection model to explain axial segregation effects in high pressure discharge lamps.

Radial demixing of reactive gases for incandescent lamps was described quantitatively in Ref. (16), where transports due to concentration and thermal diffusion in an infinitely long rotationally symmetric lamp were calculated. This procedure solved the second order partial differential equations assuming non - vanishing currents. Thus it was possible to achieve results for radial transport rates e.g. from the filament to the bulb.

The effect of strong variations of concentrations of reactive gases in incandescent lamps from the filament to the bulb has led to the suggestion that segregation effects might be observed in axial direction, as was done before in discharge lamps. Therefore the radial diffusion model was extended to a three dimensional model (17). By assuming rotational symmetry for a linear cylindrical lamp, where no convection effects had to be taken into account, the number of variables could be reduced so that the problem became solvable. Details of the mathematics of this routine have been published in Ref. (18).

This model allows to predict sites of deposition of compounds at the bulb; as an example the deposition of tungsten bromide at the bulb of an cylindrical incandescent lamp filled only with bromine and krypton (18) could be experimentally verified. Of course we are even more interested in the tungsten transport rates of the halogen cycle to determine the time of burn out of the filament. Therefore the results achieved with this model for the commonly used CH2Br2 filling with a small impurity of oxygen are presented here.

The filament temperature were taken from a measured distribution along the coil of a lamp of the type 6369R (Fig. 6). The operating voltage of this lamp is 30V, and its power consumption 250W. The maximum temperature in the middle of the filament has been measured to be 3170K. The bulb temperature was assumed to be 1000 K everywhere. The temperature field was calculated before solving the transport equations. The filling of this lamp (cold pressure) was 10 mbar CH2Br2 in addition to 5 bar Kr and an impurity of 0.05 mbar O2.

For this chemical situation the model predicts the reduction of the carbon contents of the gas phase due to the formation of tungsten carbide (W2C) at the cold leads of the filament to 7.e-5 bar; thus carbon is nearly lost as a getter for gaseous oxygen. It can be found that reactive halogen can be found mainly near the quartz bulb.

In Fig. 2 the local distribution of the total tungsten current is shown which has been calculated for this filling. The currents are calculated at the small crosses, length and direction of the lines give magnitude and direction of the current density. One finds that tunsgten is carried away from hot parts of the filaments and deposited again near the cold leads. Due to the geometry of this lamp the highest transport rates do not occur at the hottest part of the filament. In Fig. 3 the total oxygen current in this lamp is shown. Obviously it can be found that a regenerative cycle is maintained by this reactive gas contributing its part to the transport rate of tungsten. The life time of of this lamp can be estimated from transport rates to be about 900 hours.

A repetition of this calculation with a lower amount of oxygen (0.5e-5 bar) shows, that tungsten transport rates can be reduced by a factor of 2.5, which emphasizes the important influence of the oxygen in this halogen lamp upon tungsten transport and life time. Further examples of results were given in Refs. (5) and (17).

III. THE FUTURE OF COMPUTER AIDED CHEMISTRY FOR LAMPS

The presented examples show, that computer modeling of high temperature chemistry may help to understand the complex reactions and transports, which take place in lamps. The application as a tool for computer aided design of lamp chemistry is dependent on good numerical algorithms and a reliable software. Its results are only as good as the thermochemical data are. Therefore further work on the evaluation of thermo-chemical experiments has to be done.

And, above all, our models are far from being complete. Further improvements have to be included to describe transports more realistically before the lamp chemistry can be designed by the aid of a computer. This means, that the contribution of convection will have to be modeled in a different way than it has been done until now, as our lamps are not infinitely long and as we have regions of recirculating flows.

Further aspects, such as kinetic effects, which have already been considered by some authors (19)-(20), may become important for our models if low temperatures regions are considered. This implies that sufficient and reliable kinetic coefficients will become available.

REFERENCES

1. B. Kopelman, K.A. van Wormer, Illum. Engng. 63, 176, (1968)
2. B. Kopelman, K.A. van Wormer, Illum. Engng. 64, 230, (1969)
3. L.N. Yannopoulos, A. Pebler, J. Appl. Phys. 42, 858, (1971)
4. L.N. Yannopoulos, A. Pebler, J. Appl. Phys. 43, 2435, (1972)
5. E. Schnedler, Computer Modelling of High Temperature Lamp Chemistry in: E.G. Zubler Ed., Proceedings of the syposium on the science and technology of high temperature light sources High Temperature Lamp Chemistry, Toronto, Ontario, Canada, 1985 The Electrochemical Society, Pennington NJ, Proc. Vol. 85-2 pages 95-119
6. T. Geszti, Acta Techn. Hung. 78, 463, (1974)
7. W.J. van den Hoek, G. Rouweler, Philips Res. Repts. 31, 23, (1976)
8. E.H. Passmore, G.L. Duggan, J.F. Waymouth, Journ IES, 195, July (1977)
9. H.S. Spacil, K.L. Luthra, Equlibrium Chemistry in Tungsten Halogen Lamps, in: E.G. Zubler Ed., Proceedings of the syposium on the science and technology of high temperature light sources High Temperature Lamp Chemistry, Toronto, Ontario, Canada, 1985 The Electrochemical Society, Pennington NJ, Proc. Vol. 85-2 pages 120-129
10. E. Fischer, Prevention of wall blackening in metal halide lamps ibid pages 47-56
11. W. van Erk, T. Rietveld, Transport of silicon and silica in rare-earth metal halide lamps, ibid pages 57-71
12. E. Fischer, J. Appl. Phys. 45, 3365, (1974)
13. E. Fischer, J. Appl. Phys. 47, 2954, (1976)
14. H.P. Stormberg, J. Appl. Phys. 51, 1963, (1980)
15. H.P. Stormberg, J. Appl. Phys. 52, 3233, (1981)
16. E. Schnedler, Philips J. Res. 38, 236, (1983)
17. E. Schnedler, Paper presented at the Third International Symposium on the Science and Technology of Light Sources, Toulouse France (unpublished) April 18 - 21, (1983)
18. E. Schnedler, High Temperature Science 19, 237 (1985)
19. E.G. Zubler, J. Phys. Chem. 74, 2479, (1970)
20. F.J. Harvey, Met. Trans. 7A, 1167, (1976)

FIGURE CAPTIONS

Fig.1a Partial pressure in a HPI-lamp filled with 20 bar Hg, 2 bar NaI, 0.1 bar TlI, and 0.01 bar InI. An impurity of 1.e-4 bar H20 and 1.e-3bar O2 was assumed. The temperature of the coldest spot is 1100 K.

Fig.1b Partial pressure in a HPI-lamp filled with 20 bar Hg, 2 bar NaI, 0.1 bar TlI, and 0.01 bar InI. An impurity of 1.e-4 bar H20 and 1.e-3bar O2 was assumed. The temperature of the coldest spot is 1100 K.

Fig.1c Partial pressure in a HPI-lamp filled with 20 bar Hg, 2 bar NaI, 0.1 bar TlI, and 0.01 bar InI. An impurity of 1.e-4 bar H20 and 1.e-3bar O2 was assumed. The temperature of the coldest spot is 1100 K.

Fig.1d Tungsten solubility in a HPI-lamp filled with 20 bar Hg, 2 bar NaI, 0.1 bar TlI, and 0.01 bar InI. An impurity of 1.e-4 bar H20 and 1.e-3bar O2 was assumed. The temperature of the coldest spot is 1100 K.

Fig.2 Local distribution of the total tungsten current density. The currents have been calculated at the places of the small crosses. Length and direction of the lines give the magnitude and direction of the current density. Filling 10 mbar CH2Br2, 0.05 mbar H2O, 5 bar Kr (cold).

Fig.3 Local distribution of the total oxygen current density. Filling 10 mbar CH2Br2, 0.05 mbar H2O, 5 bar Kr (cold).

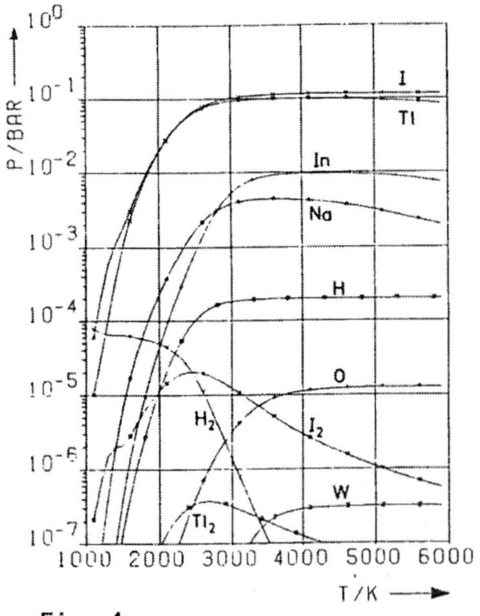

Fig. 1a

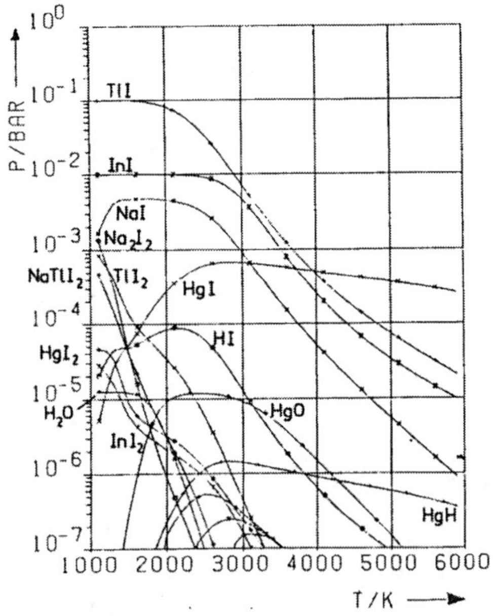

Fig. 1b

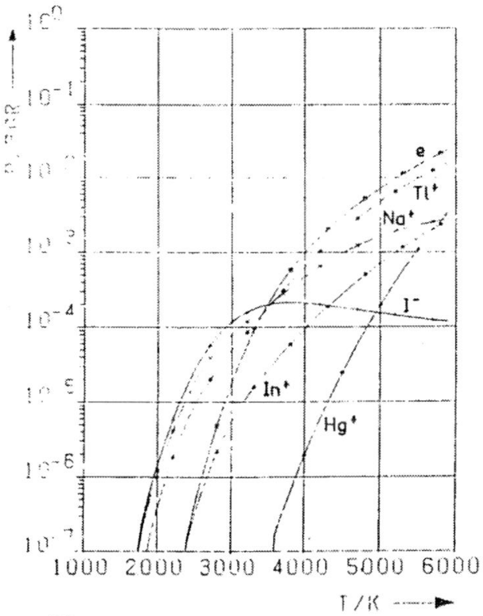

Fig. 1c

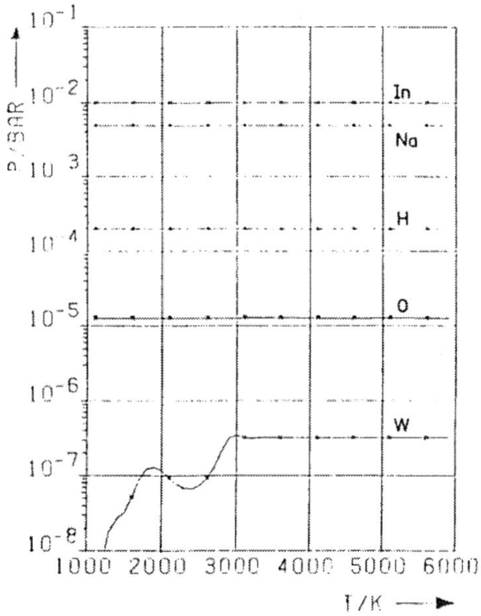

Fig. 1d

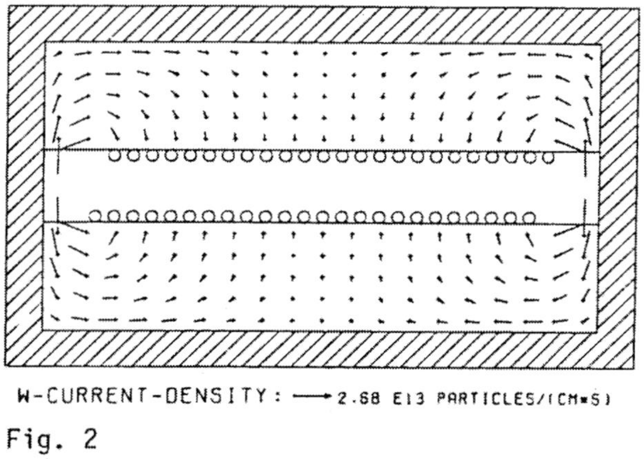

Fig. 2

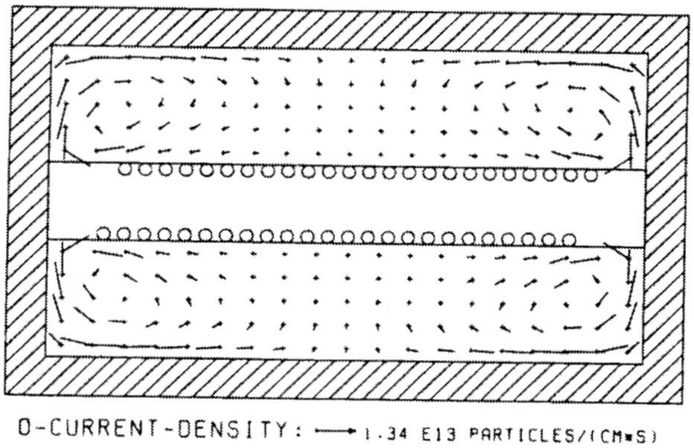

Fig. 3

C O N T R I B U T E D   P A P E R S

HIGH PRESSURE LAMPS
WITH METAL HALIDES

1 - 15

POLYNOMIAL APPROXIMATION OF THE CONDUCTANCE OF A HIGH PRESSURE MERCURY DISCHARGE . APPLICATION TO ELECTRICAL CIRCUITS FOR HP.Hg LAMPS.

J.J.DAMELINCOURT, M. AUBES and T. ABDENNADHER
Laboratoire de décharges dans les gaz UA 277 du CNRS, CPAT
118 route de Narbonne, 31062 TOULOUSE CEDEX (France)

M. STAMBOULI and M. ANNABI
Ecole Normale Supérieure de l'Enseignement Technique
5 avenue Tana Hussein, TUNIS, Tunisie

The time dependence of the lamp conductance may be written

$$\frac{dG}{dt} = \frac{d}{dt}\left(\frac{g}{h'}\right) = \frac{1}{h'}\int_0^R 2\pi r \frac{d\sigma}{dT}\frac{dT}{dt}\,dr \qquad (1)$$

Where σ is the electric conductivity and h' the effective length of the discharge. If the assumption is made that the shape ξ of the temperature profile

$$T(r,t) = T_c(t) .\ \xi\left(\frac{r}{R}\right) \qquad (2)$$

is time independent, the energy balance equation reduces to :

$$\rho C p \frac{dT_c}{dt} = \sigma E^2 - \bar{\varepsilon}(T_c) - q(T_c) \qquad (3)$$

Where $\bar{\varepsilon}(T_c)$ and $q(T_c)$ are the radiative and thermal losses by unit volume. Then :

$$\frac{dG}{dt} = \frac{E^2}{h'Cp}\int_0^R 2\pi r \frac{d\sigma}{dt}\frac{\sigma}{\rho}\,dr - \frac{1}{h'Cp}\int_0^R 2\pi r \frac{d\sigma}{dT}\frac{\bar{\varepsilon}(T_c)+q(T_c)}{\rho}\,dr \qquad (4)$$

for a time independent profile T_c=g(G) and the material functions may be calculated as functions of G ; then :

$$\frac{dG}{dt} = \frac{f(G)}{G^2}\, i^2 - h(G)$$

In a rough modelling of the discharge, with a constant pressure p, a channel-profile of time independent reduced radius β and a channel temperature T_c, f(G) and h(G) write :

$$f(G) \equiv \frac{2}{5p}\frac{1}{\pi\beta^2R^2h'}\frac{eV_i}{2kT_c}G^2 \ ;\ h(G) \equiv \frac{2}{5p}\frac{1}{h'C_p}\frac{eV_i}{2k}G\frac{\bar{\varepsilon}(T_c)+q(T_c)}{T_c} \qquad (5)$$

with :

$$\bar{\varepsilon} = \frac{B_p}{kT}\exp\frac{-eV}{kT_c} \ ;\ q = \frac{1}{\pi\beta^2R^2}\frac{2\pi}{\ell n\ 1/3}\left(a(T_c-T_w) + \frac{b}{2}(T_c^2-T_w^2)\right)$$

V_i=10.40 V ; B=1.08 10^{-10} W ; $\bar{V}$=7.8 eV ; $\beta \cong 0.5$; $a \cong 10^{-2}$ W $m^{-1}T^{-1}$

b = 8.6 10^{-6} W $m^{-1}T^{-2}$

Figure (1) gives the variation of $f^*(G)$ with G^{*2} and of $h^*(G)/G^*$ with G^* for $5300 \leq T_c \leq 6500$ where f^*, h^*, G^* are the variable part of f, h, G. Taking into account the results of figure (1) one can write :

$$f^*(G) \times G^{*-2} = a' \quad ; \quad h^*(G) \times G^{*-1} = b_3' G^{*2} + b_2' G^* + b_1'$$

and we finally obtain the time derivative of G under the form :

$$\frac{dG}{dt} = ai^2 + b_3 G^3 + b_2 G^2 + b_1 G$$

The values of parameters a and b_i have been determinated, using a recursive method, from the a.c arc voltage characteristic of a 400W HP Hg lamp and applied to lamp circuits. Figure 2 gives, as an example, the variation of conductance of such a lamp.

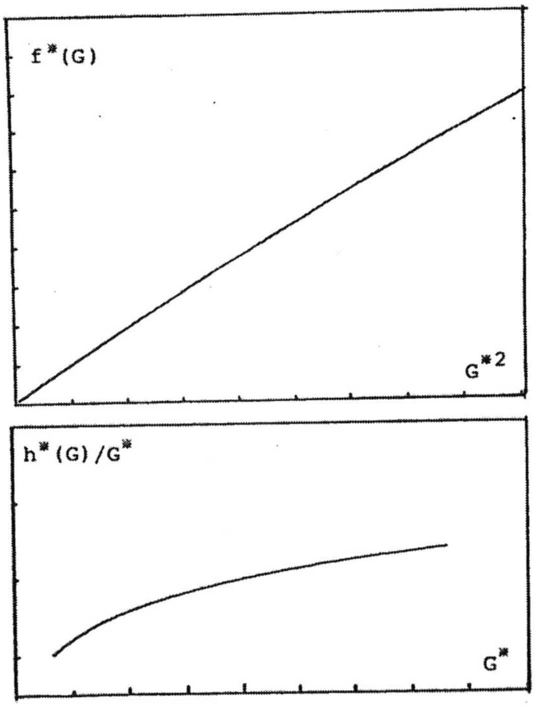

Fig.1-Variation of $f^*(G)$ and $h^*(G)/G^*$

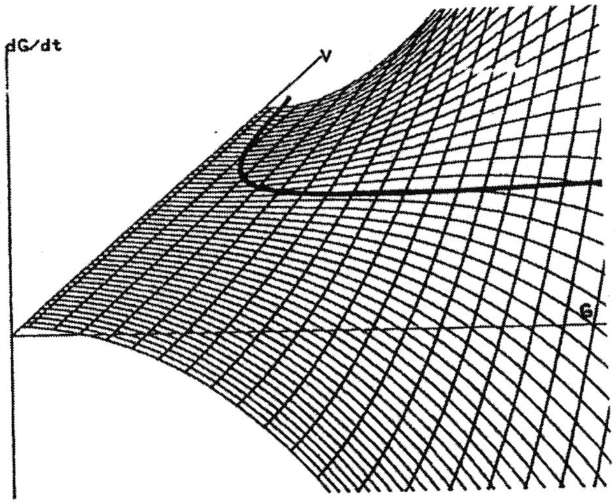

Fig.2-Variation of (dG/dt) versus G and V.

$a = 2.8$; $b_3 = 7.53\ 10^5$; $b_2 = 4.03\ 10^4$
$b_1 = 1.41\ 10^2$

Acknowledgment : we wish to thank NV Philips Industrie Fabrieken de Turnout which partly provided the lamps used in this study.

GEOMETRICAL PROPERTIES OF LUMINANCE DISTRIBUTION APPLICATION TO CALCULATION OF THE NET EMISSION COEFFICIENT

J.J. DAMELINCOURT, M. AUBES, B. BELLON
Laboratoire des décharges dans les gaz (UA 277 du CNRS) CPAT,
118 route de Narbonne, 31062 TOULOUSE CEDEX, France,

E. ROAUX
Service recherche, Compagnie des Lampes et du Matériel d'Eclairage
LIMEIL BREVANNES, France.

In LTE situation the radiative transfer equation writes :

$$\frac{\partial L}{\partial s} = \varepsilon - KL = K\ (B-L) \qquad (1)$$

if the discharge is infinite in length and cylindricaly symetrical the radiative flux reduces to its radial component and the net emission coefficient U^* reduces to $\frac{1}{r}\ \frac{\partial}{\partial r}\ (rF_R)$

$$\text{where } F_R = \int_{\psi=0}^{\pi/2}\int_{\theta=0}^{\pi/2} (L^+(r,\theta,\psi) - L^-(r,\theta,\psi))\cos^2\theta\ \cos\psi\ d\theta\ d\psi \qquad (2)$$

L^+ and L^- are the luminance at point r of beams of light of directions θ, ψ, flowing from the boundary of the discharge to point r in the r^+ and r^- directions. In the following we denote :

$$f(r,\theta,\psi) = L^+(r,\theta,\psi) - L^-(r,\theta,\psi)\ ; \qquad (3)$$

$$L'_x(r,\theta,\psi) = \frac{\partial}{\partial x} L(r,\theta,\psi)\ ;\ L''_x(r,\theta,\psi) = \frac{\partial^2}{\partial x^2} L(r,\theta,\psi)\ ;\ x=r,\theta,\psi \qquad (4)$$

The quantity $f(r,\theta,\psi)$ equals zero for $\theta=\pi/2$ and $f(r,\theta,\psi)\cos^2\theta\cos\psi$ rapidely decreases when θ or ψ goes to $\pi/2$. So, and exact knowledge of $f(r,\theta,\psi)$ is only required for directions close to θ and ψequal to zero in derivating F_R.

The next idea was to fit the exact surface $f(r,\theta,\psi)$ by an approximate one $g(r,\theta,\psi)$. This surface being defined so as its radius of curvature are equal to that of the exact surface $f(r,\theta,\psi)$ in the vicinity of θ and ψ equal to zero.

The study of $L(r,\theta,\psi)$ has shown it is possible to compute F_R

$$F_R = 4\int_0^{\pi/2} g(r,\theta,\psi)\ \cos^2\theta\ \cos\psi\ d\theta\ d\psi \qquad (5)$$

with

$$g(r,\theta,\psi) = f(r,0,0)\cos\psi\ \cos\mu\psi\ \ \cos\lambda\theta$$

μ and λ are obtained by equating the radius of curvature of $g(r,\theta,\psi)$ and $f(r,\theta,\psi)$ as before said.

The values of $L(r,0,0)$, $L''_\theta(r,0,0)$ and $L''_\psi(r,0,0)$ required in derivating the values of λ and μ are obtained from solution of (1), taking into account the symetries of the system and the properties of the derivatives of L.

The preceding calculations apply for a monochromatic radiation. In calculating the total net emission coefficient we have written U^*, for Hg discharges under the form :

$$U^* = U_{abs} + U_{na} \quad \text{with } U_{na} = p_{(Atm)} \frac{A_o}{kT} \exp - \frac{e\overline{V}}{kT} \ ; \ \overline{V} = 7.7\text{eV} \ ; \ \frac{A_o}{k} = 7 \ 10^{10} \quad (*)$$

U_{abs} is obtained from the preceeding modelling with :

$$\frac{\varepsilon}{K} = A_1 \ p_{(atm)} \exp(-\frac{eV_e}{kT}) \ ; \ K = \frac{A_2}{T} \ ; \ A_1 = 80(*) ; V_e = 4.75 \text{ eV} \ ; \ A_2 = 1.3 \ 10^6 \quad (*)$$

For the six experimental situations studied the actual results follow those obtained from the Stormberg's formulation (1) within a margin of error less than 15%.

Acknowledgement : This work, partially supported by "Club EDF Lumière et Vision" is a part of the joint program between CPAT/UA 277 du CNRS and "Service recherche et éclairage de la Compagnie des Lampes et du matériel d'éclairage". We wish to thank the "Centre Industriel de Pont-à-Mousson" for technical support.

---

(1) STORMBERG and SCHÄFER, J. Appl. Phys. 54(8) August 1983.

(*) $\frac{A_o}{k}$ (W/cm^3 Atm K) ; A_1 (W/cm^2 Str Atm) ; A_2 (cm^{-1} K)

ARC MODEL AND DIAGNOSTICS OF THE HIGH PRESSURE MERCURY ARC WITH SODIUM IODIDE ADDITIVE

James T. Dakin
Corporate Research and Development
General Electric Company
Schenectady, New York 12301
Richard P. Gilliard
Lighting Business Group
General Electric Company
Nela Park
Cleveland, Ohio 44112

Spectroscopic data from a high pressure mercury arc with sodium iodide additives are compared with model calculations. The fused quartz arc tube is oriented horizontally, and rotated about its own axis at sufficiently high speed to center the arc.[1] This mode of operation suppresses convective effects, creating a greater degree of axial symmetry and longitudinal homogeneity than is normally present in a metal halide arc. The suppression of convective effects, and the less complex spectrum and chemistry make this arc more amenable to modeling than are more conventional metal halide arcs.[2,3]

High resolution line of sight spectra are recorded for all of the principal spectral features and for a series of offsets from the arc axis. Several independent measurements of the radial arc temperature profile are extracted from the spectra using standard techniques. The temperature profile and information from the sodium lines are used to determine the radial variation in atomic sodium density.

The arc model assumes local thermodynamic equilibrium, axial symmetry, longitudinal homogeneity, absence of convection, and time independence. Equations associated with energy conservation, radiation transfer, chemical equilibrium, and multi-component dissusion are solved simultaneously to determine a self-consistent arc temperature profile, radiation field, and distribution of chemical species. The radiation transfer model includes 17 mercury lines and 28 sodium lines, most of which are treated as optically thick, and various continuum mechanisms. Boundary conditions on these calculations include the input power per unit length, the arc tube radius, the amount of mercury per unit length, and the sodium iodide vapor pressure at the wall.

Direct comparisons between the data and the model are made for the high resolution spectra at various offsets, the arc temperature profile, and the sodium density profile. Comparisons are also made between these data and data for a similar arc without sodium iodide.

[1] C. Kenty, J. Appl. Phys., vol. 9, pp. 53-66 (1938).

[2] R.J. Zollweg, R.W. Lieberman and D.K. McLain, J. Appl. Phys., vol. 52, pp. 3293-3303 (1981).

[3] H.P. Stormberg, J. Appl. Phys., vol. 52, pp. 3233-3237 (1981).

MODEL OF AC GLOW-TO-THERMIONIC-ARC TRANSITION IN DISCHARGE LAMPS

John F. Waymouth
GTE Lighting Products
Sylvania Lighting Center
Danvers, Mass.

An elementary mathematical model has been developed to describe the transition from glow discharge (with high cathode fall) to thermionic arc (with low cathode fall) during the ignition of a discharge lamp. Ionization is assumed to have been established as a result of breakdown. The electrodes are assumed to be heated by ion bombardment on the cathode half cycle and electron bombardment on the anode half cycle. A time dependent model is assumed for the cathode glow, in which ion space charge is assumed to be produced (at a rate determined by the instantaneous cathode fall) by electrons emitted from the cathode. The ions arrive at the cathode at some later time (determined by ion transit time, one of the important system parameters), delivering energy to the cathode equal to the instantaneous cathode fall at the time of arrival. Electrons are emitted from the cathode by secondary emission, proportional to ion current, with the value of gamma independent of cathode fall or temperature. This model results in glow discharge current decaying with time during those parts of the half cycle when applied alternating voltage is less than the steady-state maintaining voltage of the glow discharge, and increasing with time during those parts of the half cycle when the applied voltage is greater than the steady-state maintaining voltage. Peak glow current, and consequently energy delivered to the cathode, is very strongly dependent on the ratio of peak open circuit voltage to steady-state glow discharge maintaining voltage.

The time dependent discharge current equations are solved simultaneously with the time dependent circuit equations, assuming an inductive ballast with a resistive component.

As a result of energy input to the electrodes, they heat up, and if they reach sufficiently high temperature, commence to emit thermionically. If the thermionic electron emission becomes comparable to the secondary emission, then ion creation and ion bombardment heating are increased, and the rate of electrode temperature rise is accelerated until eventually the low cathode fall thermionic arc is established. Some twelve system parameters must be specified to describe the problem; results of calculations to determine the degree of influence of these on the duration of the glow phase will be reported. The most critical variable is the peak cathode glow current, which is determined by the ratio of peak open circuit voltage to steady-state maintaining voltage of the glow discharge; the latter is a function of the probability of ionization in the gas, and the coefficient of secondary electron emission of the cathode. Surprisingly, the thermionic work function of the cathode has a less important effect.

TEMPORAL BEHAVIOR OF METAL ADDITIVES IN HID LAMPS

H. L. Rothwell, Jr.
GTE Products
Salem, MA, USA

The most common mode of operation for high pressure metal halide (HID) arc lamps used in general lighting is alternating current (ac) at power line frequencies. In the United States the frequency is 60Hz while in most of Europe the frequency is 50Hz. Even though a metal halide lamp operates efficiently at these frequencies the combination of convection, electrophoresis, and several diffusion processes can produce undesirable lighting effects when the lamp is operating vertically. Dale Work (1), while examining the sodium and scandium partial pressures in a 175 watt HID lamps, demonstrated the complexity of additive segregation when the arc column operated vertically, and Zaengel (2), using a laser scattering technique demonstrated the large mobility of sodium ions during an ac cycle. Finally, Fohl (3), using an expanded midsection arc tube and Fischer (4), using a "bypass" arc tube, demonstrated the effect of convection on the distribution of metal additives in vertically operated arcs.

In this paper we report on some detailed measurements depicting the behavior of mercury, sodium, and scandium atoms in a high pressure arc during an ac operating cycle. Using a multi-detector, phase selectable polychromator(5) we have been able to monitor optically thin emissions from the metal additives and construct a picture of the material movement in the arc column during ac operation. As a typical example the measured partial pressure of sodium is shown in Figure 1 for a vertically operating arc during the two extreme electrode operating phases. Under this condition convection competes with cataphoresis to produce a nonsymmetrical distribution of the additives. When the top electrode is operating as a cathode the movement of sodium ions toward th top is limited by convection so that very little sodium reaches above the midsection of the arc tube. During the alternating half cycle cataphoretic pumping dominates to produce an enhancement of sodium near the bottom electrode. Averaged over the entire cycle the sodium emission becomes concentrated in the lower half producing an effect called additive segregation. In Figure 2 the same plot is presented for a vertically operating arc contained in a "shaped" arc capsule. For this situation an increase in the axial convection in the bottom half of the column drives the metal additives upward, augmenting the cataphoresis. During the cathode down cycle convection prevents cataphoretic pumping from dominating.

We have investigated both the temporal and spatial behavior of sodium and scandium additives for horizontally and vertically operating arcs at 50 and 60 hz operation. From these measurements we have developed a shaped arc tube that clearly demonstrates the effects of convection on additive segregation and consequently greatly reduces undesirable light "flicker" which occurs to sodium-scandium HID lamps when operating vertically.

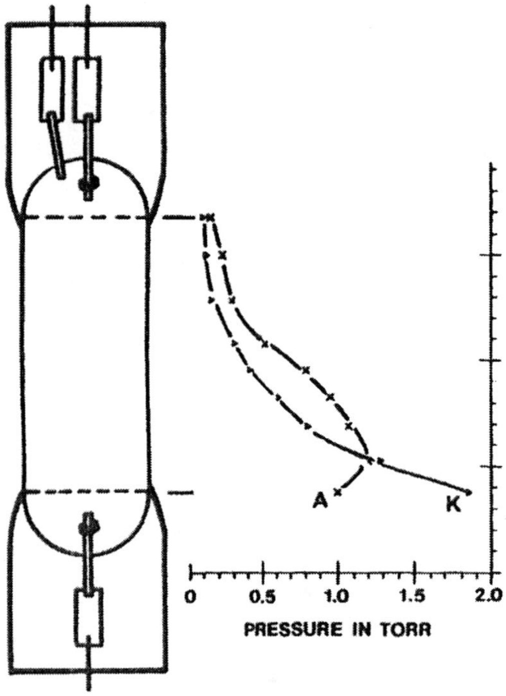

Fig. 1 - Variation in Na Partial pressure for vertical arc during peak anode (a) and cathode (K) cycles of the bottom electrode. Standard tube.

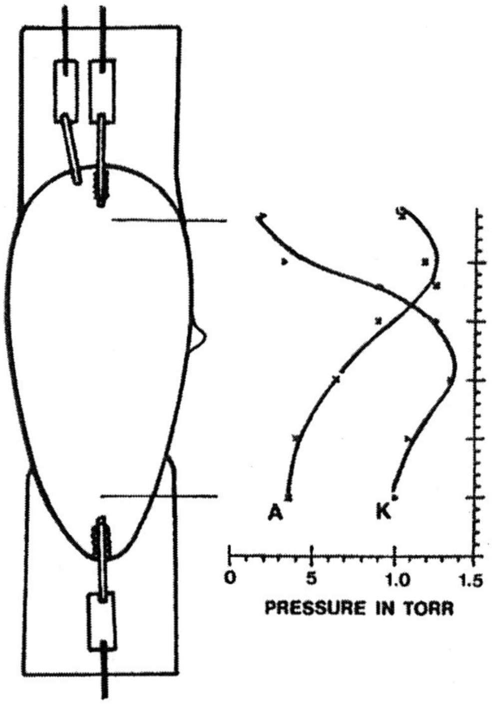

Fig. 2. - Variation in Na Partial pressure for vertical arc during peak anode (A) and cathode (K) cycles of the bottom electrode. "Shaped" tube, as in U.S. Patent 4,540,373.

1. Dale E. Work, Axial species densities in 175-watt sodium scandium metal halide lamps, Journal of IES, January 1978.
2. Th. Zaengel and E. Groiss, Laser scattering from metal halide arcs, Third International Symposium on the Science and Technology of Light Sources, Toulouse, France 1983.
3. T. Fohl, Convective effects in arcs, Journal of IES, July 1975
4. E. Fischer, J. Appl. Phys. 45,3368(1974).
5. H.L.Rothwell, Jr. and W.M.Keeffe, A direct reading polychromator for real time diagnostics of HID lamps, Journal of IES, October 1980.

SODIUM LOSS IN NaI CONTAINING METAL-HALIDE LAMPS

F.Nagel, J.Szántó, J.Ádám and I.Péter

TUNGSRAM Ltd.Co. H-1340 Budapest Váci út 77. Hungary

We have reported on the ionic mechanism of sodium loss in M-H lamps at the conference of the Electrochemical Society (Toronto, 1985).

After the discussion we have reexamined the possible presence of neutral Na, the influence of UV photoelectrons and thermal electron emission of the support rods on the sodium loss.

1. In spite of increasing the relative sensitivity to neutral particles by using deflector plates for primary ions, no neutral Na atoms could be detected, so the ratio of Na^+/Na is $>$ 100.
2. No neutral Na atoms could be found between the burner and the outer wall by atomic absorption. The Na-sensitivity was changed by a Na-filled Knudsen-cell. This arrangement enabled the detection of a $Na^+/Na > 1000$ ratio.
3. Similarly to the research team of Sylvania burnt experimental lamps were cut into segments. After etching the inner surface to remove the possible additives, the segments were dissolved and their Na-content was analysed by atomic absorption. All the segments and the original silica-tube had the same sodium content, about 14 ug/g. Furthermore no positive deviation was found in the vicinity of the support rod.
4. The surface of a similar burner was tested also by SEM-microprobe. No significant sodium peak was found by rotating the burner.
5. We prepared and burnt a series of 250 W sofitte lamp, having fused silica outer bulb and containing only a small amount of NaI.

a. normal lamps for comparison,

b. lamps with built-in 01.5 Mo wires connected to one end of the lamp in the outer bulb , modelling the support rod,

c. the same as b. but irradiated during burning by strong UV,

d. similar to b. with gold-plated Mo wire.

Type a. showed only little change in the first 100 hours.
Type b. and c. and d. have lost the same amount of their Na content in the first 100 hours and black spot occured on the wall along the wire.
All these prove that neither UV photoelectrons nor thermal electrons have significant effect on sodium loss.

6. Considering the capture cross section and the mean residence time of Na^+ shows, in agreement with experimental evidence, that photo- and thermal electrons have no significant probability of neutralizing Na^+ ions on the surface of the burner or in the space of the outer bulb.

OPERATION AND EFFICACY MEASUREMENT OF ELECTRODELESS HID LAMPS

John M. Anderson
General Electric Co.
Corporate Research and Development
Schenectady, NY 12301

The divergence-free electric field necessary to start and maintain electrodeless discharges is obtained from a changing magnetic field. If skin depth reduces penetration of the alternating magnetic field into the plasma, maintenance can be had by energy deposited near the surface as, for example, in the well-known plasma torch. If skin depth is much greater than the discharge dimensions, a ring discharge forms such that the cross section of the ring resembles closely the positive column of low frequency or dc arcs. This simplifies analysis considerably and permits more nearly a comparison with conventional arcs between electrodes.

At a frequency of 13.56 MHz, skin depth for conventional arc columns (dia. ~ 1 cm) in mercury vapor (3 atmospheres) is about 3 cms. A discharge with input power 400-1000 watts was established inside a water-cooled solenoid, 8 turns with diameter 7 cms, and height 5 cms. The discharge ring, established in a sealed quartz envelope of pillbox shape, had a major diameter of about 3.5 cms and a minor diameter of 0.6 cm. Arc loading was 30-100 watts/cm and voltage gradient 12-20 volts/cm. The solenoid was resonated by a variable vacuum capacitor and matched to the 50 ohm line from the power source by a capacitive divider. To start, the vapor was overvolted, and capacitors adjusted as needed during lamp warm-up.

Power input must be measured accurately. This was done by digitally acquiring the coaxial line voltage and current and multiplying in an oscilloscope. Pickoff of line voltage and current was by a capacitor voltage probe and a current transformer. As a basic power standard, a 60 Hz reference standard wattmeter, basic traceable to the National Bureau of Standards, was employed. A water-cooled resistive coaxial load was used at 60 Hz and the low frequency wattmeter agreed within one percent with energy loss calculated from input and output load water temperature. The load was then used at 13.56 MHz and the oscilloscope calibrated at the high frequency. The scheme correctly finds net lamp power even with lamp and line mismatch. The test lamp and standard lamp were located at balanced positions in a 0.9 meter light-integrating sphere and alternately operated, both remaining in the sphere for all measurements. Standardized lamps of the high pressure sodium and metal halide types were traceable to the National Bureau of Standards.

Losses in the circuit between the power source and lamp discharge must be measured so that net power into the plasma is accurately known. Loss at 13.56 Hz in the solenoid (~50-100 watts) was measured by the differential temperature of the cooling water for the solenoid. Circuit losses (~10-30 watts) in the matching network were approximately determined by the difference between input power and solenoid loss without lamp ignition. An overall computer program based on an equivalent circuit, was written to calculate all voltages, currents, and impedances which depended upon input measurables and solenoid current. Stray capacities were included. A further loss in the measurement is light blockage by the solenoid. This was determined empirically by alternately placing a small incandescent source within and outside the solenoid while measuring integrated light. The results of measurements with electrodeless lamps is presented in a companion paper.

P.D. Johnson, C.E. Jones, and T.H. Rautenberg all of this Laboratory, collaborated in this work. The effort was supported by the US Department of Energy, San Francisco Office, through the Lawrence Berkeley Laboratory of the University of California, and is described in Report LBL-19035 dated January, 1985.

EFFICACY MEASUREMENTS FOR UNCONVENTIONAL FILLS IN ELECTRODELESS HID LAMPS

John M. Anderson

Peter D. Johnson

General Electric Co.

Corporate Research and Development

Schenectady, NY 12301

Operation of HID lamps without electrodes, described in a companion paper, allows fill ingredients which may give improved efficacy, but which unfavorably react with the electrode material to limit lamp life. In this work we have measured the efficacies of electrodeless lamps using conventional metal halide fills to establish the procedure, and then extended measurements to include the bromides and chlorides of promising atomic radiators. The latter are known to react with varying severity with electrodes. Molecular radiators, principally mercury chloride, were also considered. Mercury was always used as a buffer gas, and small additions of argon helped starting.

Lamp arc tubes were constructed of quartz in the shape of a pillbox with outer diameter 5.3 cms, height 2 cm and wall thickness 2-3 mm. The arc tube was generally encased in a thin layer of quartz wool for insulation, and inserted into a larger pillbox with removable cover. This was in turn placed in the exciting solenoid. When hard starting was encountered, the ingredients were partially condensed with liquid nitrogen which quickly evaporated in operation.

A survey of promising elements, and those already tried in lamps, was made, and most of the periodic table eliminated. Left, as molecular radiators, were the chlorides and bromides of mercury, tin, copper, and ytterbium as well as the chlorine molecule.

Atomic radiators judged to be of promise were sodium, thallium, cerium, ytterbium, cesium, and scandium.

Five lamps with conventional fills were constructed. Mercury at 3 atmospheres and 75 watts/cm loading gave a positive column efficacy of 63 lm/W, slightly higher than the positive column efficacy of electroded lamps. Thallium, as the iodide, gave about 85 lm/W. Sodium and thallium, with or without scandium, gave about 135 lm/W, all at about a loading of 85 watts/cm. Arc column voltage gradients were comparable with electroded lamps.

Twenty-one lamps were constructed with more unconventional fills, seven intended as molecular radiators and the rest atomic. As a general comment, the bromides and chlorides as molecular radiators gave low efficacy, 50 - 80 lm/W, and high color temperature. They were sometimes unstable in operation. Of the numerous combinations of atomic radiators tried, noted above, the best result was obtained with a fill of TlBr, $CeCl_3$, and NaBr, with mercury. This lamp had a positive column efficacy of 170 lm/W, color temperature of 4700 K, CRI of 51, and C.I.E. chromaticity $x = 0.37$ and $y = .45$. No obvious band structure was noted, but cerium provided a strong, almost continuous, spectrum of atomic lines to enhance color.

C.E. Jones, T.H. Rautenberg, and J.R. Young, all of this Laboratory, collaborated in these measurements. The effort was supported by the US Department of Energy, San Francisco Office, and will be described in report No. DOE/SEF/11994-1. Probable date of issue will be January, 1986.

High frequency operation of HID lamps

Eugen Statnic

OSRAM GmbH
D-8000 München 90

Introduction

High operating frequencies are interesting for small high intensity discharge lamps (HID) because very compact electronic ballasts with low power losses can be used by increasing system eficacy and versatility of the lighting systems. When operated with higher frequencies, however, all high pressure lamps are subject to arc instabilities that are caused by acoustic resonances.

Acoustic resonances are intolerable for the operation because the instabilities are usually associated with changing of the color temperature, lower luminous efficacy and optical disturbances such as annoying light fluctuations, fluttering and flickering.

Advantages of high frequency operation

Higher luminous efficiency (3...10 %), about 200...300 K lower color temperature, arc without typical "banana" convection curvature, higher light density, up to 15 % higher system efficacy and compact electronic ballast are relevant advantages. Furthermore a better maintenance should be noted: only 12 % flux drop after 6000 h, lower rise of lamp voltage,negligible shift of color temperature, and reduced end blackening due to electrode evaporation.

The electronic ballast holds the lamp power constant for new and old lamps by compensating the voltage lamp rise influence on the lamp power during the life.

The "hot-restart" of the lamps in each phase of cooling is possible with simple circuitry only in high frequency mode of operation.

Resonances in HID lamps

As known before, the high frequency periodic input power heats the discharge gas, and produces periodic pressure oscillations, and under certain conditions even standing pressure waves are possible. The resonances are due to these standing pressure waves when the power input is in phase with the pressure wave. The sound velocity C in the discharge and the arc length L_B are determining the basic longitudinal resonance frequency of each type of HID lamp: $f_{RL} = C/2L_B$. Radial, azimuthal or diagonal basic resonancies are possible too. The frequencies of basic resonancies f_R depends on the geometry of the arc tube interior and the sound speed distribution in several plans of the fluid flow.

In fact, the frequency of the power input is twice the applied current frequency i.e. the resonance driving frequency is always half of resonance frequency: $1/2\ f_{RL}$.

Similar to organ pipes in lamps different arc instabilities are relatively strong at the first three or four integral harmonics of all basic resonances. But intolerable instabilities could also be excited by a power input of lower frequency which had large Fourier components at the basic resonance frequency. Otherwise, if the energy of lamp current harmonics is high enough to excite a standing wave, an arc or at least an aureole instability can occur at 1/6, 1/4 and 1/3 of all basic resonance frequencies.

Especially the metal halide lamps due to several ion species show a particularly large number of resonance frequencies resp. resonance bands. As a result, a high frequency operation is possible in principle only in some distinct windows between resonance bands but also outside i.e. below first resonance or above all resonances. We find experimentally that each type of lamp, has a range free of resonance below $1/6\ f_{RL}$ and another range above approximately $4\ f_{RL}$ (fourth integral of the basic longitudinal resonance). By modulating the frequency of the high frequency supply voltage, lamp operation may be largely stabilized and thereby the light parameters (luminous flux, color, etc.).

INSTANT START MINIATURE METAL HALIDE LAMP

AKIHIRO INOUYE, KIYOKAZU HONDA
LAMP RESEARCH AND DEVELOPMENT DEPARTMENT
TOSHIBA CORPORATION
1-201-1 FUNAKOSHI YOKOSUKA 237 JAPAN

A metal halide lamp has high intensity, high luminous efficacy and long life compared with an incandescent lamp. But it needs a few minuits to become full light output after ignited.

This limits the wide uses and applications of the metal halide lamp. Therfore the instant start miniature metal halide lamp has been developed for extending the metal halide lamp applications.

The miniature metal halide lamp has an ellipsoidal arc tube of which the short diameter is 6mm and the long diameter is 10mm. The electrodes gap is 5mm. The consumption power of this arc tube is 35 watts.

To realize instant starting ability, the following technology is applied.

(1) An ceramic heater, with 30 watt operating power, has been mounted adjacently to the arc tube. The arc tube is pre-heated before a lamp starts or is heated up immediately as a lamp starts to gain light out put rapidly.

(2) New electrodes of the discharge tube have been developed to endure high current input during lamp starting.

(3) Pulse wave form is composed with two parts. One is very high voltage and narrow pulse to break-through the arc gap. The other is low voltage and wide pulse to build up the arc discharge between the electrodes.

(4) Radio isotope is added into the arc tube because of producing initial electrons which enable the electrode gap to breakdown instantly.

Thus, the metal halide lamp has been realized to start or re-ignite immediately by these means after the lamp is switched on.

This lamp can emit the light over 75% of the rated light output immediately after switched on. This can be usefull for video camera lights or vehicle headlamps.

NEW DEVELOPMENT OF METAL HALIDE LAMPS FOR PHOTOGRAPHIC PURPOSES

Bernd Lewandowski
OSRAM GmbH, Hellabrunner Str. 1, 8000 Munich 90
Federal Republic of Germany

Metal halide lamps for TV, movie and stage illumination (HMI) are presented and their properties described.
HMI-lamps are characterized by a mainly continuous multi-line spectrum which is emitted by means of the high multiplicity of the atomic levels of the rare earths (dysprosium, holmium, gadolinium) and the appearance of multiple excitation of electrons.
HMI-lamps are distinguished especially by light-electric energy conversion (90 lm/W) in connection with good colour properties (corrected colour temperature CCT: 5600 K or 6000 K, colour rendering index ($R_I \geq 90$) which allows an efficient generation of daylight.
As it became necessary to develop daylight colour lamps with higher luminous flux, the HMI-lamp family (HMI 200 W-HMI 6000 W) was completed by a 12 KW-lamp. In order to decrease the temperature of the bases, all HMI-lamps, except the HMI 200 W and HMI 575 W, were manufactured with frosted stems which prevents the dominant light-pipe effect by multiple total reflections.
Furthermore, with the HMI 2500 W/GS (gap shortened) a lamp is presented which, respecting the electrical and mechanical compatibility, in comparision to HMI 2500 W shows an increase of the luminous flux of about 10 % and of the mean luminance of about 30 %; this effect, during a spot regulation of the luminaire enters the overall efficiency of the system lamp-luminaire directly. It should be noted that the short-arc-technology will be transferred to the HMI 1200 W and HMI 575 W types.
Another step of development is the introduction of the single-ended technic with respect to the HMI 250 W/SE. Compared with the HMI 200 W the light from a luminaire is improved distinctly because of the axially mounting of the lamp in the reflector and a reduction of the electrode gap by a factor of 2.

OPTIMIZATION OF 100W MH LAMP DESIGN

By

W. M. Keeffe and Z. K. Krasko

GTE Lighting Products

Sylvania Lighting Center

Danvers, Massachusetts 01923 USA

A new 100 watt metal halide lamp was developed, featuring high efficacy of 85 lpw, a warm, 3200K, color temperature with a color rendering index of 65. The excellent color was obtained without a use of phosphor coating. The lamp design was optimized through the extensive experimental investigation of radiometric, colorimetric, thermal and electrical characteristics along with their maintenance through lamp life.

The main design features of the new lamp are:

1. Elevated arc tube wall loading of $15.5 w/cm^2$.
2. Optimized metal halide stoichiometry with $NaI:ScI_3=24:1$.
3. Optimized electrode insertion length and "backspace" between the rear of the electrode coil and the point of entry into the arc tube press.
4. Tubular quartz sleeve closed at one end, surrounding the arc tube.
5. Compact size, ED17 vacuum outer jacket.
6. Improved arc tube mount design.

The design parameters described have been fine-tuned in our final development to achieve the desired lamp characteristics. The spectral power distribution of the lamp is full, well balanced and rich in red. An extremely high density of scandium lines extend throughout the visible spectrum, improving considerably color rendering (Fig.1). The rich red portion of the spectrum is due to the assymetrically highly broadened sodium resonance line and a manifold of strong scandium lines at 620-630nm. From an absolute power calibration, the visible radiation is found to be 33.7W for 100W input (Fig.2). Infrared radiation is measured at 15W and UV is 5W (about half of UV for higher lamp wattages). Tables I, II and III present photometric, electric and physical characteristics of the new M100W lamp.

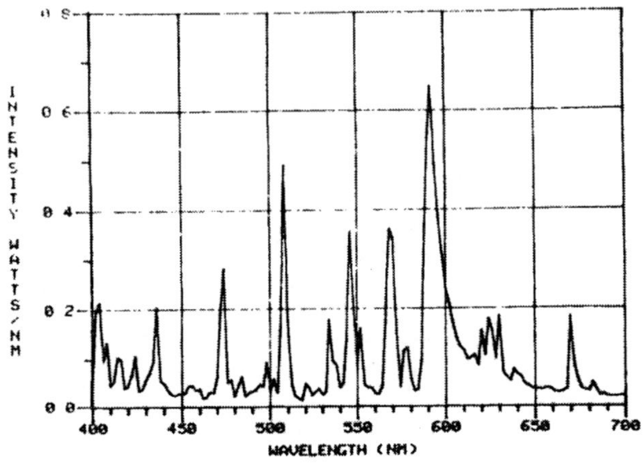

FIGURE 1

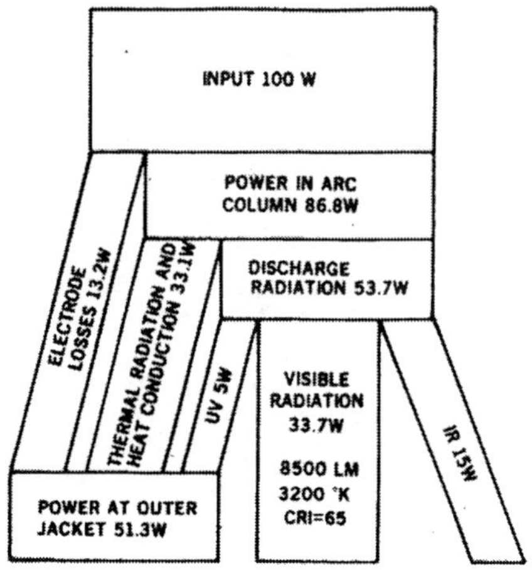

FIGURE 2

TABLE I

Photometric Characteristics

Initial (100 Hours) Lumens Vertical Operation	8500
Mean Lumens (40% Life)	6800
Average Rated Life (Hours)	5000
Correlated Color Temp. °K	3200
Color Rendering Index	65
Warm-up Time (Minutes)	2
Hot Restart Time (Minutes)	5-10
Chromaticity Coordinates X =	.420
Y =	.395

TABLE II

Electrical Characteristics

Nominal Lamp Watts	100
Nominal Lamp Volts	100
Nominal Lamp Amps	1.1
Min. Starting Volts:(See Note (2.)	
98% Probability of Starting at 0°F or Higher (Sine Wave)	220
90% Probability of Starting Down to - 20°F	220

TABLE III

Physical Characteristics

Base Designation	Medium Screw Base
Bulb Material	Borosilicate (Hard Glass)
Bulb Designation	ED17
Nominal Bulb Diameter	54MM (2 1/8")
Maximum Overall Length	138MM (5.43")
Light Center Length	86MM (3.39")
Burning Position	Vertical (See Note)
Maximum Bulb Temperature (°C)	527 F (275°C)
Maximum Base Temperature (°C)	374 F (190°C)
Eccentricity (Base to Bulb	3°
Eccentricity (Base to Arc Tube)	3°
Bulb Finish	Clear

Absents

~~Nr.~~ 13

Residual Gases in the Outer Jacket of Metal Halide Lamps

Zhu Yu Hua, Wu Chong Ruo, Yuan Ci and Wang Hai Ou
Department of Electronic Engineering
Nanjing Institute of Technology, Nanjing, China

The discharge performance, luminous efficiency, rendering colour and operating life of metal halide lamp are affected significantly by the dangerous gaseous impurities within the arc tube. The hydrogen is the most harmful for all gaseous impurities. The presence of hydrogen makes to rise both the starting and re-ignition voltage of the lamp. Simultaneously, the dangerous residual or impurity gases within the glass outer jacket also markedly affect on the operating characteristic of these lamps. During lamp operation, the wall of arc tube being at high operating temperature, the hydrogen easily permeates into the arc tube. It is obvious that studying the gases in the vacuum jacket is extremely important. If the hydrogen partial pressure within the jacket is much higher than in the arc tube, some of the hydrogen diffuses through quartz wall into arc tube, and influences the discharge performance. Generally, a getter material is put in the proper place of the jacket, so releasing gaseous impurities from the quartz tube, glass wall and other components will be sorbed by the getter.

In this tests, the glass jacket is connected to a B-A vacuum gauge and a omegatron. The former is used to measure the total pressure, and the latter is used to analyze the residual gases in the jacket. Omegatron being especially suitable to measure gases at static state, the variation of the residual gases in the jacket is obtained by this method.

After the lamp is burned for about 40 hours, the total pressure within a jacket without any getter raised to about 10^{-1} Pa. Therefore a Ba-Al getter have been mounted in it. Before flashing barium getter, the main residual gases are $CO(N_2)$, H_2, CH_4; after flashing the partial pressure of oxygenated gases, such as CO, CO_2, H_2O, is reduced evidently; H_2 and He are almost unchanged;

but CH_4 and Ar are increased relatively much. Since barium element is very active, it reacts easily with oxygenated gases; whereas the hydrogen is difficultly absorbed by Ba; CH_4 and Ar are released from the getter material. If a non-evaporable getter Zr-Al is employed mounting on near two ends of the arc tube, it can efficiently absorb the hydrogen. As a result, the hydrogen partial pressure in the jacket is less than that only that the Ba-Al getter is used by a factor of ten.

The relationship of the residual gases variation with the lamp life is shown in this test. During the life tests, the H_2, He and Ar are greatly increased, because H_2 is released from lamp components; the He is permeable through glass from air; the Ar arise from the desorption of the gas previously sorbed by the getter, but the CH_4, CO and CO_2 will be gradually reduced.

Finally, in order to understand for us the effect of hydrogen on the performance of the lamp, the jacket is filled with H_2 up to 1 Pa, the experiments have shown that the starting and the ends voltage of the lamp almost rised linearly, it is concluded that H_2 has continuously gone into arc tube.

Iodure de fer / Iodure de Cobalt / → UVA (315-400 nm)

Nr. 14

A SEARCH FOR AN IMPROVED ULTRAVIOLET RADIATION SOURCE

UVC : 200-280
UVB : 280-315
UVA : 315-400 -

R B Page, THORN EMI Lighting Limited,
Research & Engineering Laboratories, Melton Road,
Leicester, LE4 7PD, U.K.

INTRODUCTION

For many years ultraviolet radiation sources, have been used in specialised areas such as photoprint exposure, suntan and water sterilisation applications. Recent developments in materials and processes have made extra demands on the required spectral response. To meet this need, gallium and/or iron iodides have been combined with mercury to produce discharge lamps with a broader, more adaptable spectrum (Ref.1). Despite these improvements, ongoing developments in the printing industry have led to film coatings which require precise exposure wavelengths whilst at the same time, the ever increasing use of U.V. sources for ink curing and water sterilization have created a demand for new lamps giving improved spectral performance.

EXPERIMENTAL

In an attempt to find additives which will enhance the spectra of existing U.V. sources, this work compares a number of metal iodides in conjunction with mercury and also in selected combination.

A series of test lamps were manufactured, using an arc tube with a fixed volume, which was successively dosed with the iodides of cobalt, manganese and antimony. Also added to each arc tube was a quantity of mercury selected to give a nominal lamp volt drop together with a fixed pressure of inert gas. Each lamp was operated at 1000W for a fixed period of time prior to an assessment of the resultant radiation response. Spectral energy measurements were made within the wavelength range 200 to 400n.m., in 5nm steps, using a U.V. spectroradiometer. The energy in the UVA, UVB and UVC spectral regions was then summed to give a relative energy performance number corresponding to each test lamp.

RESULTS

These can be divided into 3 categories:

a) Effect of increasing dose quantity on spectral performance

From previous work on gallium and iron iodides, (Ref.1) it was expected that a peak U.V.A radiation output would be achieved at a particular iodide dose. In the case of manganese this was found to be substantially true.

Cobalt was found to behave in an unexpected manner. Initially the U.V.A. output decreased with increasing iodide dose and then reversed to reach the highest level so far achieved.

The UVB and UVC output progressively reduced with increasing iodide quantity in all of the materials under test. Although the peak values were not accurately located the results strongly indicate that dose values around 1mg will give the best response.

b) Iodide response at specific wavelengths

As expected, each of the selected iodides radiated strongly at specific wavelengths. Cobalt proved to be of prime interest, generating radiation at 340 to 360nm and 240 to 280nm, which are spectral areas of particular importance for the suntan and water sterilization applications respectively. In these regions manganese and antimony were less efficient. Manganese however, was found to be the best UVB (280 to 315nm) radiation generator of the group.

c) Spectral response of selected iodide combination

Whilst each of the trial iodides could be used to produce a useful lamp for a particular application a favourable combination of iodides would have a much wider market appeal. It was therefore decided to assess an iron iodide - cobalt iodide - mercury system. The result show that a wider coverage of the UVA spectrum can be achieved using this iodide combination without undue loss of energy at specific wavelengths. The UVB and UVC spectral regions were not enhanced over the performance which could be achieved by mercury alone.

CONCLUSIONS

Cobalt iodide was shown to be a promising additive to conventional mercury discharge sources which, in certain proportions, can enhance both the UVA and UVC spectral regions.

A selective combination of cobalt and iron iodides together with mercury can produce a broader wavelength response in the UVA region than can be achieved by a single iodide additive alone. Manganese iodide was found to be the most effective UVB generator of the test group.

REFERENCE

R B Page - 3rd Int. Symp. of Light Sources, Toulouse, April 1983

INVESTIGATION OF DOPING POSSIBILITIES OF RARE GAS FLASH LAMPS

M. Neiger, Chr. Popp and H.-P. Popp
Lichttechnisches Institut, University of Karlsruhe, Kaiserstr. 12
Federal Republic of Germany

Dedicated to Prof. Dr. P. Schulz on the occasion of his 75th birthday

I. Introduction

Rare gas flashlamps, especially Xenon filled ones, are powerful sources of pulsed light. Contrary to e.g. photographic illumination applications, their broadband, gray-body spectral emission is a disadvantage for such applications as photolithographic illumination or optical pumping of lasers, where only spectral emission into the corresponding absorption band is required. Doping of rare gas flashlamps is a possible means to achieve such a goal. We have studied such doping to achieve more selective emission in the UV range 250 nm - 300 nm.

II. Experimental Approach and Results

The study was performed with fast pulse ($\sim$10 - 20 μs) flashlamps with electrical energy inputs of 0.1 - 1 kJ or specific electrical energy loadings of 5 - 50 J/cm^3 . To maintain the high spectral radiance of short pulse flashlamps, the high plasma temperature of 15 000 - 20 000 K has to be maintained. Since most dopants have relatively low ionization energies, the line radiation of ionized elements seems most suitable. The second column of the periodic table with ns^2 - valence configuration is promising because of very strong resonance lines in the first ionic spectra (alkaline-like spectra). Ionic dopant radiation has the additional advantage of very little self-absorption in cold boundary layers near the wall. With these considerations and the desired spectral range of 250 nm - 300 nm Magnesium seemed to be a promising dopant. To avoid blacking of the lamp wall by metallic Mg-layers, either MgF_2 or MgO were introduced as fine powder into the lamp bodies.

Experiments with different rare gas fillings showed that Ne is the best choice for a high dopant concentration in the plasma. Preconditioning the lamp by creating a dopant aerosol before each shot is most effective. Fig. 1 shows time-integrated flashlamp spectra in the selected UV range. The low continuous background, desirable for good selective emission, is obvious. The spectral radiances are comparable to those of high intensity short pulse Xenon flash lamps in the same spectral range.

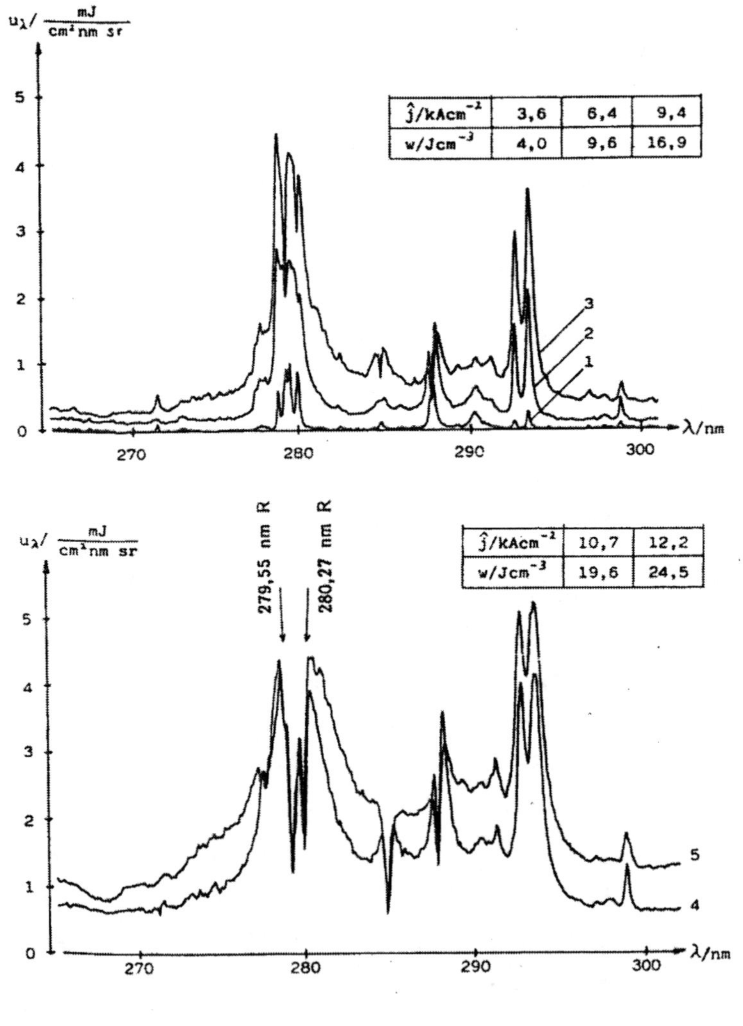

Fig. 1: Time integrated spectra of Mg-doped flashlamp at differnt specific energy loadings.
(Numbers 1 - 5 represent rising energy inputs.)

HIGH PRESSURE LAMPS WITH SODIUM AND XENON

16 - 24

THE INFLUENCE OF NaNa, NaHg AND NaXe MOLECULES ON THE SPECTRUM OF THE HIGH-PRESSURE SODIUM LAMP

J.J. de Groot, J.P. Woerdman* and M.F.M. de Kieviet

Nederlandse Philips Bedrijven B.V., Lighting Division, Eindhoven, The Netherlands
* Huygens Laboratory, University of Leiden, The Netherlands

The visible and near infrared spectrum of the standard high-pressure sodium (HPS) lamp is dominated by the self-reversed sodium D-lines with some structure (bands) in the very far wings. For the calculation of the spectrum [1] the broadening of the D-lines is usually described in terms of resonance broadening which is, strictly speaking, only valid for the impact broadening regime near line centre, where the long-range dipole-dipole Na-Na interaction is dominant. In recent years the insight has grown [2] that the very far wings of the D-lines are in fact of (quasi-)molecular origin, originating from short range Na-Na interaction. Similarly the buffer gases mercury and xenon influence the wings of the D-lines, because of Na-Hg and Na-Xe interaction.

In the quasi-static approximation [2] one general expression is obtained for the absorption coefficient of bound molecules and quasi-molecules, which can be evaluated when the potential energy curves of the relevant levels as functions of the internuclear separation of the atoms are known. Starting from the known potential curves of the NaNa, NaHg and NaXe molecules the absorption coefficients have been calculated for the various transitions as a function of wavelength and temperature. Maxima in the absorption coefficients (satellites) are found which correspond with the experimentally observed bands for Na (551, 805, 882 nm), Na+Hg (671 nm) and Na+Xe (560 nm) discharges. Combining the absorption coefficients for the molecular and the Na atomic transitions the emission spectrum of the HPS lamp can be calculated with the help of the radiative transfer equation [1]. Including the molecular contributions gives a good agreement with measured spectra as shown in figure 1 for an HPS lamp with a relatively high sodium vapour pressure. Numerical data will be presented which show the influence of the (quasi-)molecules on light technical data as chromaticity coordinates, colour rendering, luminous efficiency of visible radiation etc.

[1] J.J. de Groot and J.A.J.M. van Vliet, J.Phys.D: Appl.Phys., 8, 651 (1975); I.E.E. Proc.Pt.A, 128, 415 (1981).

[2] J.P. Woerdman and J.J. de Groot, Chem.Phys.Lett., 80, 220 (1980); ACS Symposium Series, no. 179, 33 (1982); J.Chem.Phys., 76, 5653 (1982); J.Phys.B:At. Mol. Phys. 18, 4205 (1985).

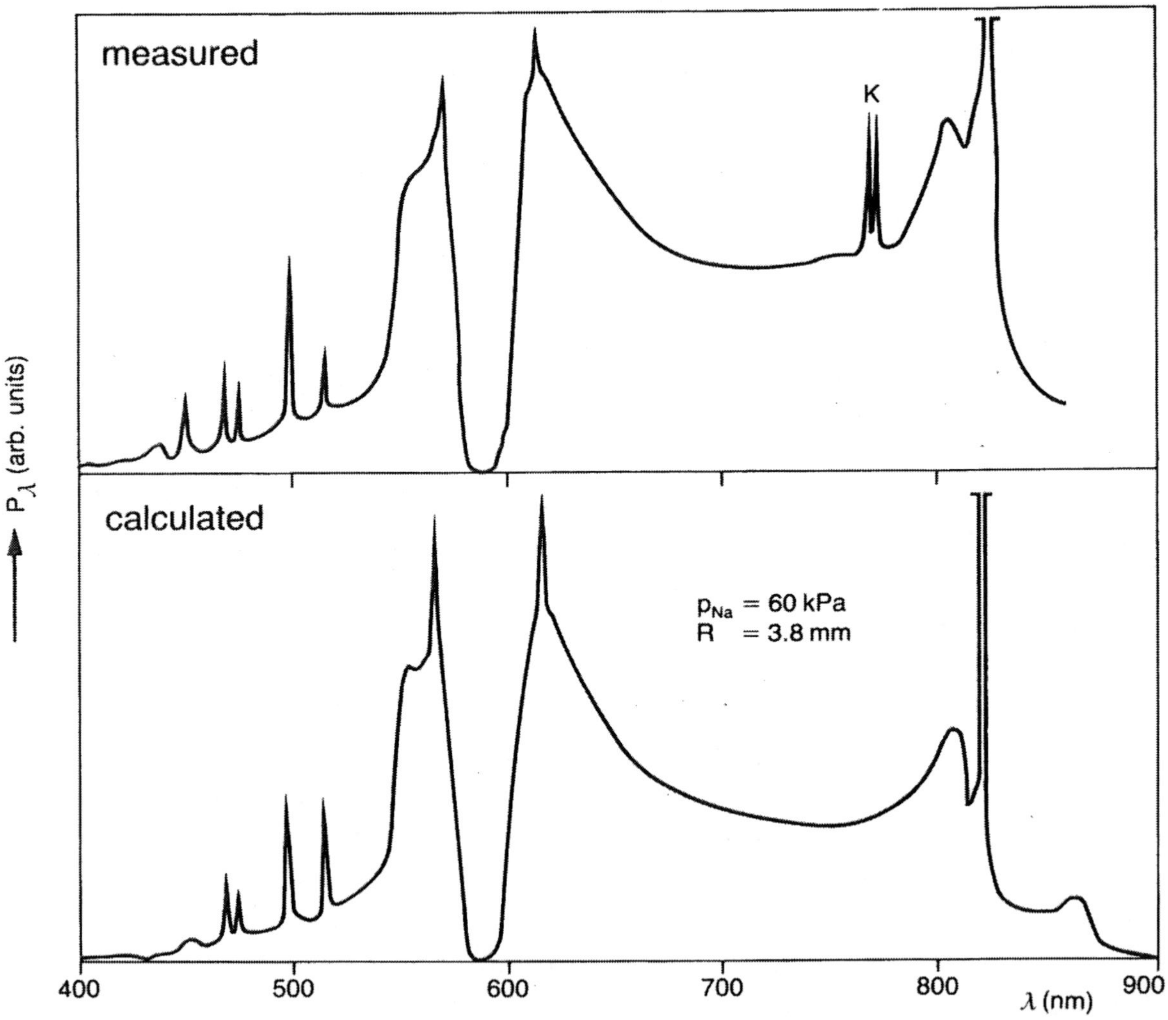

Fig. 1: Comparison between measured spectrum and calculated spectrum (including NaNa contributions) for a "pure" high-pressure sodium discharge.

LOW-WATTAGE HPS LAMPS FOR SPECIAL USE AND TESTS MADE ON THEM

Gy.Székács and D.Vida , Tungsram Co.Ltd., Light Source Development Dept., H 1340 Budapest, Váci út 77. Hungary

Favourable features of HPS lamps are,for the most part utilized in public lighting applications. Their wattage range presently expands 35 to 1,000 watts. Improvements in technology have made it possible to fabricate lamps with wattages below 35 watts. This wattage range below 35 watts gives the possibility for special applications differring from the present ones. This special field of application includes e.g. vehicle lighting mobile machine lighting and hand lanterns in which lamps are mounted in optical systems suitable for shaping narrow beam of light.
For these purposes HPS lamps have not been used until now.
These applications have required the development of special lamp constructions.
Constructions problems of HPS lamps for these purposes and tests performed on practical lamps made under experimental conditions are discussed.
These include:

- Reduction of rated wattage to the 16 to 35 watts range;
- Reduction of arc tube dimensions in order to achieve an approx. 10-mm arc length;
- Improvements in the arc tube heat balance by using special electrode configurations and niobium lead-in wires;
- Provision for instant starting and restarting;
- Shortening of stabilization time compared to that of "standard" HPS lamps presently used in public lighting.
- Optical tests made on the arc tube:
  - transparency of the arc tube wall;
  - finding a relationship between apparent arc dimensions and focussing ability;
  - tests made to establish relationship between colour temperature, colour rendering and luminous efficacy;
  - mechanical vibration resistance tests;
  - temperature measurements.

The discharge tube of our experimental lamps was made of sintered alumina tube with an electrode spacing of approx. 10 mm. High in-line transmission characteristics of the ceramic discharge tube enables good focussing.
The electrodes of special configuration are welded to niobium wires making electric conduction across the sintered alumina end plugs. Using this construction, a favourable heat balance of the discharge tube end has been achieved which made it possible to develop unusually low-wattage lamps.
Positioning compact lamps incorporating 16 to 35-W HPS discharge tubes in a rotational parabolic mirror, it was observed that the light from the discharge tube can fairly be focussed into a beam. The beam can be formed, similarly to incandescent lamps, to predetermined shapes using optical shields positioned near the discharge tube.

THE DISCUSSION OF THE CHARACTERISTICS IN SAPPHIRE HIGH PRESSURE SODIUM LAMP

Cai Zuquan : The Institute for Light Sources
Chen Dahua : at Fudan University in Shanghai of China

Yu Heging : The Institute for Building Material in Wuxi of China

ABSTRACT

Compared sapphire high pressure sodium (HPS) lamp with general HPS lamp, we have discovered that the distant illuminance of sapphire HPS lamp increases by 50 per-cent or more, that has attracted the wide attention and controversies to the illuminating scientists. It is considered that this abnormal phenomena is a result of the varying with the materials of electric arc tubes for gas discharge lamps. The traveling decrease of light wave energy is different while they go through differential materials, that results in the different characteristics between sapphire HPS lamp and general HPS lamp.

In this article, we present a calculating method for luminance contrast of HPS lamp, which basic principle is that the gas discharge lamps of sapphire HPS lamp and general HPS lamp are considered to be the ideal Lambert light sources. Although the calculating method is approximate, it may successfully be explained a series of new phenomena of sapphire HPS lamp. Obviously owing to the advantage of sapphire HPS lamps, it will predict that HPS lamp will be able to enter a new developping stage.

THE SAPPHIRE ARC TUBE AND IT'S APPRICATION IN THE HIGH PRESSURE SODIUM LAMP

Yu Heging : The Institute for Bulding Material in Wuxi

Chen Dahua : The Institute for Light Sources at Fudan University in Shanghai of China

ABSTRACT

It is described that we have adopted the advanced technology of edge-defined film-fed growth method (E.F.G.) to produce the transparant sapphire material as the envelope of high pressure sodium (HPS) lamp. The cost of the producing sapphire tube can appropriately be compared with the translucent ceramics tube. We have also introduced the characterstics of the HPS lamp with the sapphire arc tube envelope. The sapphire HPS lamp is able to increase the illuminance, to extend the actual life and to improve the color rendring, compared with the general HPS lamp in the same condition. At the end of this paper the discussion is made for the properties of sapphire HPS lamp.

DESIGNING FILTERS TO IMPROVE THE COLOUR PROPERTIES OF HIGH-PRESSURE SODIUM LAMPS

D A J Mottram, A R Vos, D O Wharmby, THORN EMI Lighting Limited, Melton Road Leicester. U.K.

Various manufacturers have, in recent years, attempted to produce improved colour versions of the high pressure sodium lamp. The resulting 'de luxe' lamps have improved colour but are still not universally acceptable for interior use. Attempts to produce a 'white-light' SON lamp have been dogged by problems of short life.

We decided to carry out a study to determine what changes to the SON spectrum were needed to give good colour properties. We ruled out additives to the spectrum as additives in SON lamps have, in the past, proved problematic. That leaves the option of subtracting radiation, that is, filtering.

The problem can now be stated as being to find that filter which achieves the colour properties with the least loss of efficacy. Mathematically, this is a case of constrained, non-linear optimization with bounded variables. We investigated various numerical methods for solving this problem, all of which require the application of computers. We found that the most successful method was a simple direct search technique based on the Hooke and Jeeves algorithm [1].

The method has been applied to the spectrum of a 'de luxe' SON lamp. This spectrum had the following properties.

Colour co-ordinates x	:	.506
y	:	.415
Colour temperature	:	2200K
R_a	:	75

We set targets for the method of a colour temperature of 2600K or greater and R_a of 80 or greater. We were also able to put constraints on the values of the CIE special colour rendering indices. The optimum filter for this target gave a predicted loss in efficacy of 22% relative to the original 'de luxe' lamp spectrum. This loss in efficacy is slightly less than that resulting from improving the colour by other methods.

REFERENCE

Himmelblau, D M, "Applied Nonlinear Programming, (New York : McGraw-Hill), 1972.

NR. 21

EXPERIMENTAL INVESTIGATION OF HIGH PRESSURE XENON- AND XENON/THALLIUMIODIDE SHORT ARC LAMPS

R. Hoppstock, M. Neiger and H.-P. Popp
Lichttechnisches Institut, University of Karlsruhe, Kaiserstr. 12
Federal Republic of Germany

Dedicated to Prof. Dr. P. Schulz on the occasion of his 75th birthday

I. Introduction

High pressure noble gas short arc lamps have an extremly high radiance, such as the well known Xenon short arc lamps. We investigate such lamps for their potential as an incoherent pump source for continuous wave dye lasers. To this purpurse the lamps are doped with additives with line radiation in the absorption band of the laser dye.

II. Details of the lamps

For our investigation we use d.c. operated high pressure Xenon short arc lamps with an electrode distance of 2 mm - 2.5 mm, similar to commercial lamps. The electrical power ratings are 150 W - 250 W corresponding to a current of 7 A - 12.5 A. The operating pressure (up to 40 bars) of the lamps is measured with a spiral manometer of fused silica which is sealed directly to the quartz envelope.

III. Temperature measurements

The high radiance and the plasma temperature of the 'hot spot' in front of the cathode tip are of special interest. The temperature measurements are made by detecting absolute radiance of the three Xenon 764 nm lines and of the Xenon continuum. The Abel-transformation from observed radiance to radial distribution of the emission of the plasma are made by the method of Bockasten /1/. The transition probabilities of the Xenon lines are calculated by the Coulomb approximation of Bates and Damgaard /2/. The continuum radiation is calculated by summing up the free-bound transitions, the free-free transitions of the electrons in the field of the ions and atoms, and a component which is related to the line radiation. This last contribution is the sum of many weak and broad Xenon lines whose superposition appears as a quasicontinuum.

These calculations which have been performed by Palacin /3/ are shown in Fig. 1. The χ -factors of the continuous emission are taken from the data of Hofsaess and Geltman /4, 5/. Results of such temperature measurements are shown in Fig. 2. The maximum of the temperature in the hot spot is approximately 11 000 K.

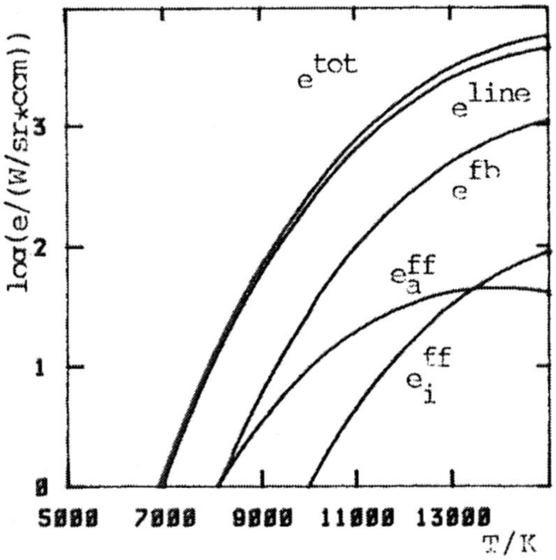

Fig. 1: Emission coefficients integrated from 725nm to 745 nm for Xe 40 bar operating pressure

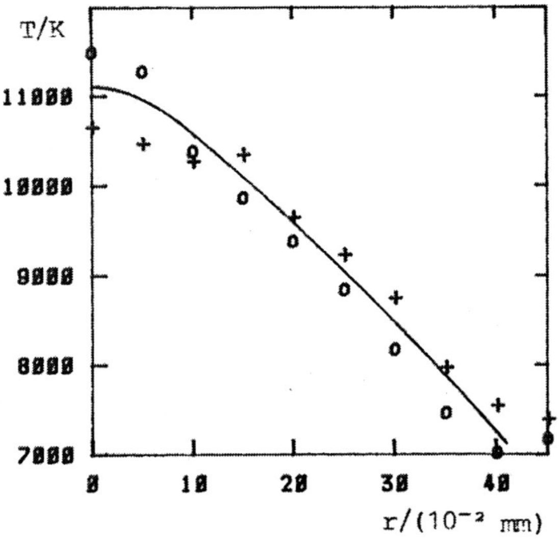

Fig. 2: Measured temperature profile of 40 bar Xe lamp, + = determined by line emission, o = determined by continuum emission

To enhance the radiance in the 500 nm - 550 nm spectral region we add 20 mg Thalliumiodide to the Xe gas. Without any external heating of the lamp bulb the temperature of the hot spot remains about 11 000 K. The density of the Tl atoms is determined from the radiance of the Tl 535-line and gives a cold spot bulb temperature of approximately 790 Kelvin. The spectrum of such a lamp compared to a pure Xe-lamp with similar operating conditions is shown in Fig. 3.

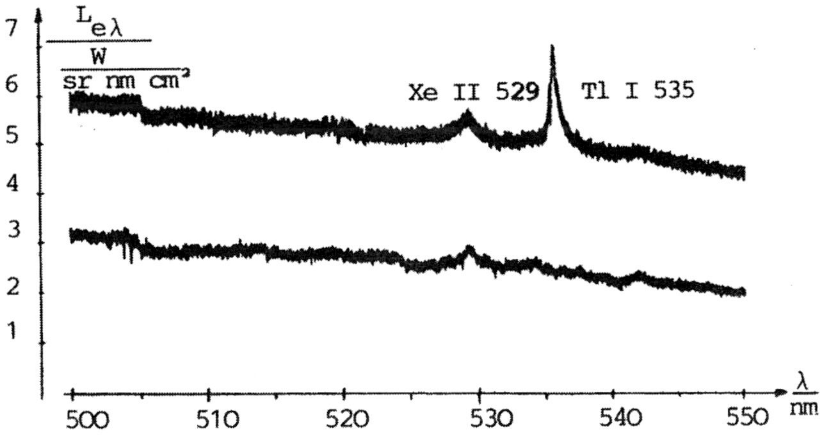

Fig. 3: Spectrum of Xe and Xe/TlI lamps

/1/ Bockasten, K.: JOSA 51, pp 943 - 947 (1961)

/2/ Bates, D., Damgaard, A.: Phil. Trans. Roy. Soc. A 242 (1949)

/3/ Palacin, A. C.: private communications

/4/ Hofsaess, D.: JQSRT 19, pp 339 - 352 (1978)

/5/ Geltman, S.: JQSRT 13, pp 601 - 613 (1973)

SIMPLE MODEL OF THE CATHODE SPOT PLASMA OF HIGH-PRESSURE SHORT ARC LAMPS

A. C. Palacin, H. Müller and M. Neiger
Lichttechnisches Institut, University of Karlsruhe, Kaiserstr. 12
Federal Republic of Germany

Dedicated to Prof. Dr. P. Schulz on the occasion of his 75th birthday

I. Introduction

The high radiation intensity of rare gas short arc lamps makes these sources suitable as e. g. incoherent laser pumps or for photolithographic purposes. To optimize the radiation intensity with respect to the absorption band of the medium in question one must know the temperature profile of the plasma. Modeling the arc behaviour requires simultaneous solutions of the set of conservation equations for energy, momentum and mass. Solutions of that problem are demonstrated for a Xenon short arc lamp with and without Thalliumiodide.

II. Modeling of the arc plasma

1. Assumptions

The model adopted for these studies is based on the following assumptions:

- (i) the arc is in local thermodynamic equilibrium (LTE),
- (ii) the arc is rotationally symmetric and stationary,
- (iii) the arc is stabilizied by forced convection due to the self magnetic field (gravity forces being neglected),
- (iv) the flow-field is laminar and viscous forces are neglected,
- (v) heat dissipation due to viscosity effect is negligible,

2. Conservation equations

The solution of the energy equation (1) determines primarily the temperature profile of the arc.

$$j^2 / \sigma = \mathrm{div}\,(\rho\, h\, \vec{v}) + \mathrm{div}\,(-k\ \mathrm{grad}\ T) + u \qquad (1)$$

where $\vec{j}$ is the current density and $\vec{v}$ is the plasma flow velocity. The arxis temperature on the axis is mainly determined by equation (1) neglecting the enthalpy flow. On the other hand the arc size is nearly completed determined by the enthalpy flow through the arc cross-section perpendicular to its axis. The plasma velocity $\vec{v}$ is given by the solution of the momentum equation (2) and mass conservation equation (3).

$$\rho\, (\vec{v}\ \mathrm{grad})\, \vec{v} = -\ \mathrm{grad}\ p + \vec{j} \times \vec{B} \qquad (2)$$

$$\mathrm{div}\,(\rho\, \vec{v}) = 0 \qquad (3)$$

The self magnetic field $\vec{B}$ is determined by Maxwell equation.

3. Material functions

The solution of the above equations necessitates the knowledge of following material functions as a function of temperature and pressure:

- (i) thermal conductivity k
- (ii) electrical conductivity σ
- (iii) mass density ρ
- (iv) specific enthalpy h
- (v) net radiation emission coefficient u

Based on the assumption of LTE it is possible to determine k, σ, ρ and h as a function of the local temperature. The calculation of the net radiation emission coefficient u is extremly complicated because it depends on the whole temperature profile. Therefore we solve the radiation transfer equation for the optical thick part of the spectrum assuming a parabolic temperature profile.

III. Results and discussion

The governing equations are solved with some crude approximations /1/ for the above mentioned short arc lamps. The lack of a realistic description of the condition at the cathode-plasma transition forces us to use the current density in the vicinity of the cathode as input parameter. The current density was determined by the estimated cross-section of the arc at the cathode region. As a typical example (Xenon short arc lamp, operating pressure p = 40 bar, I = 8 A, electrode distance 2.5 mm). Fig. 1 shows the axial temperature distribution in front of the cathode tip. The axial velocity of the plasma on axis (r = 0) is shown in Fig. 2.

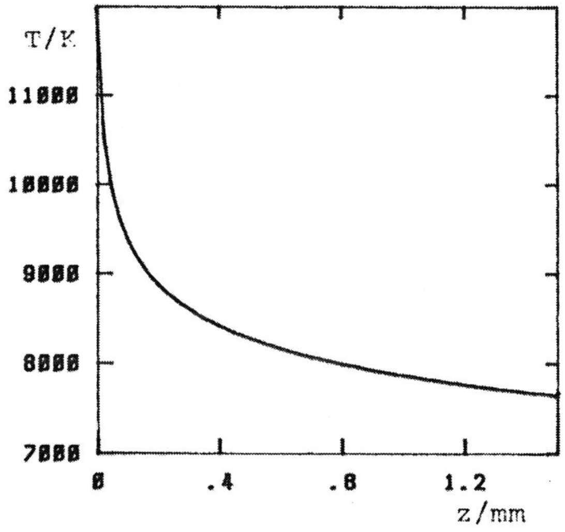

Fig. 1

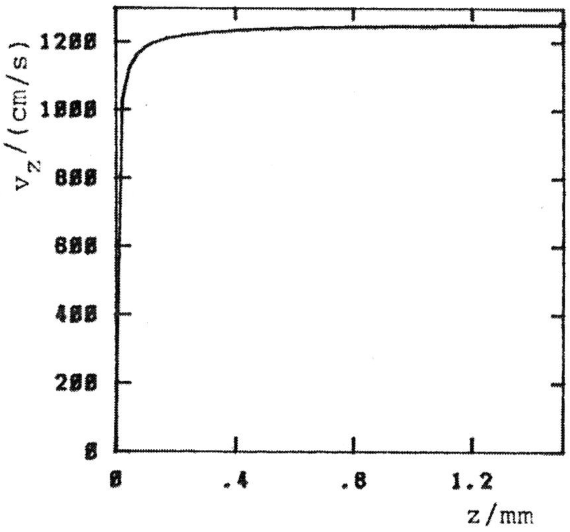

Fig. 2

/1/ Lowke, J.J.: J. Phys. D. 12, pp 1873 - 1886 (1979)

USE OF A STATISTICAL MODEL TO EVALUATE THE TOTAL RADIATIVE SPECTRAL LINE POWER FROM A THERMAL PLASMA: APPLICATION TO HIGH PRESSURE XENON LAMPS

A. C. Palacin and M. Neiger
Lichttechnisches Institut, University of Karlsruhe, Kaiserstr. 12
Federal Republic of Germany

Dedicated to Prof. Dr. P. Schulz on the occasion of his 75th birthday

I. Introduction

Recently a statistical model has been proposed to provide a relatively simple means of evaluating the total radiative spectral line power from thermal plasmas /1/. The procedure is best applicable to optically thin situations like multi-line lamp spectra (e. g. rare earth doped lamp plasmas) or to dense overlapping line spectra producing a so-called quasicontinuum (e. g. very high pressure rare gas lamp plasmas). In this paper the model is applied to high pressure Xenon lamps. All resonance lines (in the VUV part of the spectrum) are not taken into account due to complete self absorption. Results are in very good agreement to earlier computations and measurements done by Günther /2/.

II. Statistical model for the total line contribution

For optical thin lines the total emission coefficient e^{bb} for all lines of a thermal plasma is given by (1)

$$e^{bb} = \frac{1}{4\pi} \frac{n}{u} \sum_{m} g_m \, e^{-\frac{E_m}{kT}} \sum_{n<m} h\nu_{nm} A_{nm} \qquad (1)$$

where A_{nm} is the transition probability, g_m the statistical weight of the upper level, E_m the energy and ν_{nm} the frequency of the emitted line.

The particle density n and partition function u are functions of the plasma temperature. The needed transition probabilities are computed by the method of Bates and Damgaard for all allowed electric dipol transitions and from the information contained in the energy level tables of Moore. We compute nearly 5 000 transitions probabilities. For better handling and to accelerate the computation time of (1) the idea of statistical spectroscopy is used /1/. It substitutes the double sum in (1) by a double integral (2), which can be solved analytically.

$$e^{bb} = \frac{1}{4\pi} \frac{n}{u} \int dE \frac{dG}{dE} e^{-\frac{E}{kt}} \int dE' \; h\nu \frac{dG}{dE'} \bar{A} \qquad (2)$$

The statistical weight of the discrete energy levels are substituted by an energy level density dG/dE. Fig. 1 shows this procedure for the Xenon level spectrum with an exponential function.

$$dG/dE = b \; e^{-aE} \qquad (3)$$

The transition probability of a certain upper level is proportional to the number of lower levels and to a mean transition probability $\bar{A}$ which is computed from the exact values. Fig. 2 shows the contributions of the lines and of the different continuum emission coefficients to the total radiated power of a p = 40 bar Xenon plasma as a function of temperature.

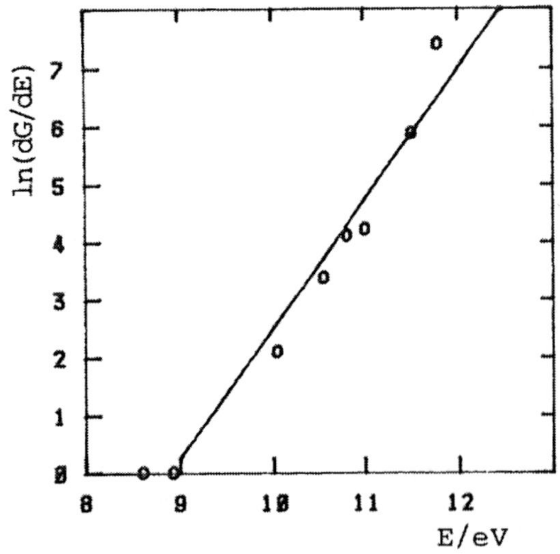

Fig. 1

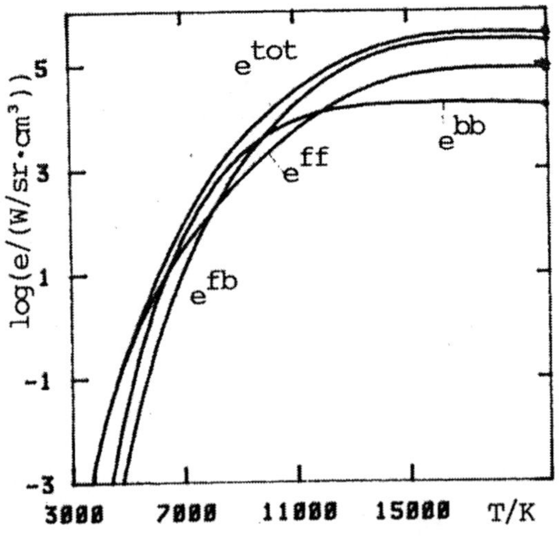

Fig. 2

/1/ Cram, L. E.: J. Phys. D 18, pp 401 - 411 (1985)
/2/ Günther, K.: J. Phys. D 12, pp 1093 - 1100 (1979)

DESIGN AND TEST OF A 10 KW XE-LAMP WITH CONSTANT INTENSITY PROFILE AT FIXED DISTANCES

Dr. K.-J. Dietz, Dr. J. Witt
W. C. Heraeus GmbH, Produktbereich Original Hanau
Heraeusstr. 12-14, D-6450 Hanau

Abstract: We report the design of a 10 kW Xe-lamp which has a nearly constant intensity profile at a given distance from the lamp. This is realized by changing the radius in the middle section of the lamp. The dimensions of the lamp and the irradiance distribution were calculated and are in good accordance with the experimental results. The arc stability was examined in the range of 35 - 530 mbar Xe pressure.

In several applications a homogeneous irradiance at a certain distance from a lamp is demanded. Usually, this problem is solved by using absorbers or special reflectors which attenuate the higher irradiance in the middle section of the light field. In our case the required distance was 24 cm, and the admitted intensity variation was less than +/- 5% over a length of 60 cm. The lamp power was approximately 10 kW.

Conventional lamps show an irradiance distribution at such distances as typically displayed in Fig. 1(a). Another possibility we have used was to optimize the irradiance distribution by changing the geometrical characteristics of the lamp as shown in Fig. 2. The middle section of the lamp is characterized by a bigger tube diameter. Physically this is described by [1] :

$$E_z \sim d^{-3/2} \quad (1)$$

where E_z is the electrical field gradient and d the diameter of the lamp tube. Thus in the middle section of the lamp the field gradient is lowered, and there is less radiation power. The optimum ratio between the two diameters and the length of the thicker part was calculated with the formula (1). The resulting calculated irradiance distribution is shown in Fig. 1(b).

Fig. 1(c) shows the measured values in good accordance with the calculation. Please note that the homogeneity of +/- 5% of the new profile is by nearly factor 2 better than that of the usual

lamps.
Sudden change in tube diameter may lead to arc instabilities, especially at higher pressures. Decreasing pressure will minimize such effects. Sufficient stability is achieved by pressures below 200 mbar. Such pressures result in decreasing UV and visible light efficiency, especially below 100 mbar (see Fig. 3 and [2]). To sum it up, the application of such a lamp is a compromise between irradiance homogeneity and light efficiency.

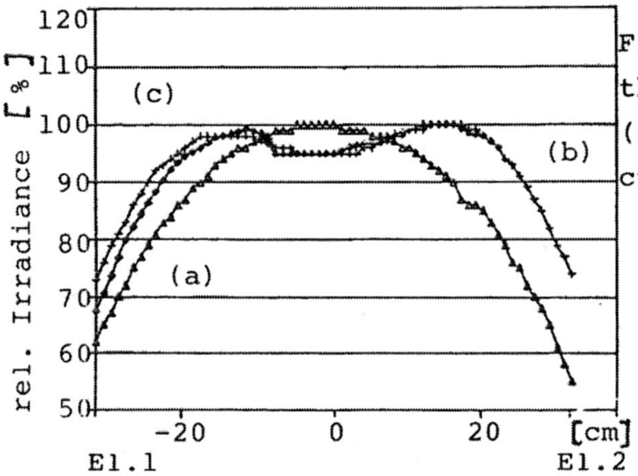

Fig.1: Relative irradiance along the lamp axis at 24 cm distance. (a) conventional lamp, (b) calculation, (c) measurement

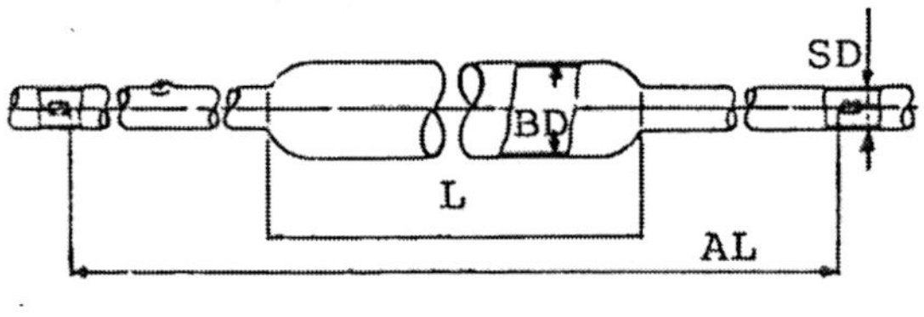

Fig. 2: Geometry of the lamp. Arc length AL = 64 cm, small diameter SD = 1.35 cm, bulb diameter BD = 2.85 cm, length of middle section L = 22 cm.

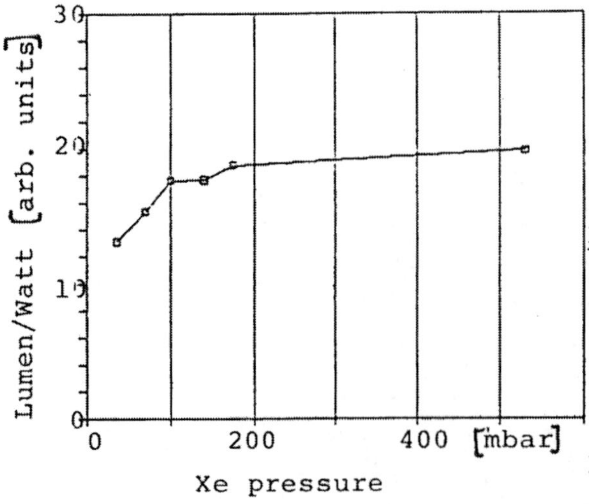

Fig. 3: Efficiency: Lumen/watt [arb.units] as a function of Xe filling pressure [mbar]

References:

[1] Elenbaas, W., Quecksilberdampfhochdrucklampen, Eindhoven 1966
[2] Weizel/Rompe, Theorie el. Lichtbögen und Funken, Leipzig 1949

LOW PRESSURE LAMPS

25 - 36

A MODEL OF RADIAL VARIATIONS IN THE FLUORESCENT LAMP POSITIVE COLUMN

James T. Dakin
Corporate Research and Development
General Electric Company
Schenectady, New York 12301

The low pressure Hg-Ar positive column has been studied with a model which accounts for radial transport mechanisms. Recent investigations of new effects, such as departure from a Maxwellian electron energy distribution and assoicative ionization, have made simplifying assumptions with regard to radial transport.[1-3] Here we explicitly treat mechanisms such as ambipolar diffusion, radial cataphoresis, thermal conduction in the electron and neutral gases, and the transport of excitation by radiation trapping. The model calculates, not assumes, the radial density distribution for each of the species and the radial temperature distributions for the electrons and the neutral gas.

The model includes ten species: Ar, $Hg(6^1S_0)$, $Hg(6^3P_0)$, $Hg(6^3P_1)$, $Hg(6^3P_2)$, $Hg(6^1P_1)$, $Hg(7^3S_1)$, Hg^+, Hg_2^+, and e^-. At each radial node, the number densities of these species are determined from rate equations. These rate equations incorporate electron impact excitation, de-excitation and ionization processes, associative processes, the emission and absorption of radiation, diffusion, and recombination. The radiation includes two resonance lines and three lines from the $Hg(7^3S_1)$ state, all treated as optically thick. Diffusion is included only for the Hg ground state and the three charged species. The temperature distribution for the neutral gas is determined by the balance between elastic electron impact heating and thermal conduction. The deviation from a Maxwellian electron energy distribution is treated using a two-temperature approach. Radial variations in the temperature of the electron gas are deduced from its thermal conductivity and radial variations in its energy balance.

Solutions are found for specified values of the wall (reservoir) temperature, electric current, Ar pressure, and tube diameter. Solutions include the above mentioned densities and temperatures plus the electric field and the radiation. These solutions are compared with available parametric data for 36 and 14 mm diameter tubes.[4-5] Cross sections for the associative processes are adjusted to optimize the agreement between the model and the data.

The shapes of the radial density distributions are studied. The electron distribution can be characterized by $\alpha = \langle n \rangle / n_e(r=0) = 0.43$. This compares to the Verweij[5] measurement of $\alpha = 0.41$, and the value $\alpha = 0.43$ from Schottky[6] theory. The calculated value of α is found to decrease with increasing Ar pressure as observed by Verweij. The depletion of ground state Hg from the axis comes from the combined effects of gas heating and radial cataphoresis. The density distributions of metastables are somewhat broader than that of the electrons.

Two different approaches to radiation trapping are considered. First, the natural decay rate for each transition at each radial node is multiplied by a Biberman-Holstein coefficient (net radiative bracket) to determine the effective decay rate. The Biberman-Holstein coefficients vary radially, fully account for excitation transport, and must be determined iteratively. Second, calculations are repeated using the standard escape factor (radial-averaged Biberman-Holstein coefficient) approach. These two approaches are found to give essentially identical results, except for slight variations in the shapes of the resonance state density distributions.

The calculations show only a 1% radial variation in the electron temperature for the standard operating condition. Larger variations are predicted for conditions with large variations in Hg or Ar density caused by gas heating or radial cataphoresis.

In conlcusion, this work supports some of the simplifying assumptions often made in modeling the fluorescent lamp discharge at the standard operating condition. Under conditions of significant gas heating or radial cataphoresis, these assumptions are less valid.

[1] L. Vriens, R.A.J. Keijser and F.A. Lighthart, J. Appl. Phys. Vol. 49, pp. 3807-3813 (1978).

[2] R.B. Winkler, J. Wilhelm and R. Winkler, Annalen der Physik, Vol. 40, pp. 89-139 (1983).

[3] R. Lagushenko and J. Maya, Journal of IES, Vol. 14, pp. 306-314 (1984).

[4] W. Verweij, Philips Research Reports Supplements, No. 2 (1961).

[5] S. L. Zhu and B.H. Zhang, Third International Symposium on the Science and Technology of Light Sources, Toulouse (1983).

[6] W. Schottky, Z. Phys. Vol 25, p. 635 (1924).

COMPUTERMODELLING OF FLUORESCENT LAMP PLASMAS

R. Sonnenburg, H. Gündel
VEB Kombinat NARVA, Forschungszentrum für Lichttechnik,
DDR 1017 Berlin, Ehrenbergstr. 11–14

A numerical model of fluorescent lamp discharges is designed on the basis of hydrodynamic balance equations for the lamp plasma and the equations for the operating circuit.

The operating circuit of a normal fluorescent lamp with a lamp ballast L and a parallel capacity C is taken into account. The lamp voltage and therefore the electric field is obtained by solving the differential equations for this circuit. In order to calculate the resistance R^* of the lamp plasma we used a three level model for the Hg-vapour including the ground state and one lumped exited level for the Hg (6^3P_0), Hg (6^3P_1) and Hg (6^3P_2) states. Further assumptions are closely related to those given in /1/. The equations for the currents I_1 and I_2 are given by (1) (2), the equations of continuity for the electron density n_e and the density of excited atoms n_m by (3) (4), the balance of the electron energy by (5) and the equation of motion of the electrons by (6).

$$\frac{dI_1}{dt} = \frac{U_0}{L}\sin\omega t - \frac{1}{CL}\int (I_1 - I_2)\,dt \qquad (1)$$

$$\frac{dI_2}{dt} = \frac{I_1 - I_2}{C R^*} \qquad (2)$$

$$\frac{\partial n_e}{\partial t} = z_{gi}\, n_g n_e - D_a \nabla^2 n_e + z_{mi}\, n_m n_e \qquad (3)$$

$$\frac{\partial n_m}{\partial t} = z_{gm}\, n_g n_e - \frac{n_m}{\tau_{eff}} - z_{mi}\, n_m n_e - z_{mg}\, n_m n_e \qquad (4)$$

$$\frac{\partial U_e}{\partial t} = \chi\, \nu_{col}(U_g - U_e) + e v_d E + \sum_{x,y} z_{xy}\, n_x \Delta U_{xy} \qquad (5)$$

$$v_d = b_e E \qquad (6)$$

The lamp resistance is given by

$$R^* = \frac{l}{\pi R_a^2 e b_e n_e}$$

By integration over the tube cross section we obtained from (1)-(5) a set of ordinary differential equations with the variable t (time). The quantities used in the above equations are the rate coefficients of the transition x→y z_{xy}, the ambipolar diffusion coefficient D_a, the electron drift velocity v_d, the electron charge e, the electric field E, electron mobility be, the electron neutral collision frequency ν_{col}, the fractional energy loss per elastic collision χ, the effective life time of the exited level τ_{eff} taken from Holstein's /2/ theory and the radius of the discharge tube R_a.

The system described above has been solved numerically using 'Euler's' method for a set of estimated values U_e, n_e, n_m, and ωt for a operating frequency of 50 Hz.

Our results for the lamp voltage and lamp current agree qualitatively with those of a standart fluorescent lamp. The existing discrepancy between the values of the calculations and the experiment are mainly due to the assumption of a Maxwell distribution function in our calculations and the uncertainties in the knowledge of the atomic datas. Fig. 1 shows the results for the calculated lamp voltage und lamp current of a half period of a 40 W lamp. The corresponding values of a standart 40 W fluorescent lamp are shown in Fig. 2. At present we perform calculations for lamps filled with Hg-Ar-Kr mixtures and others.

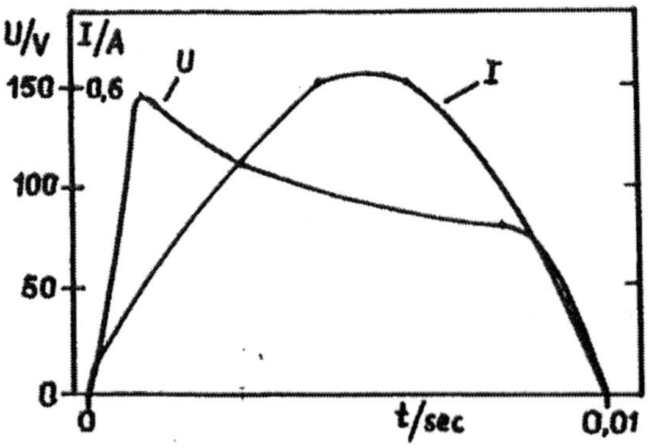

Fig. 1

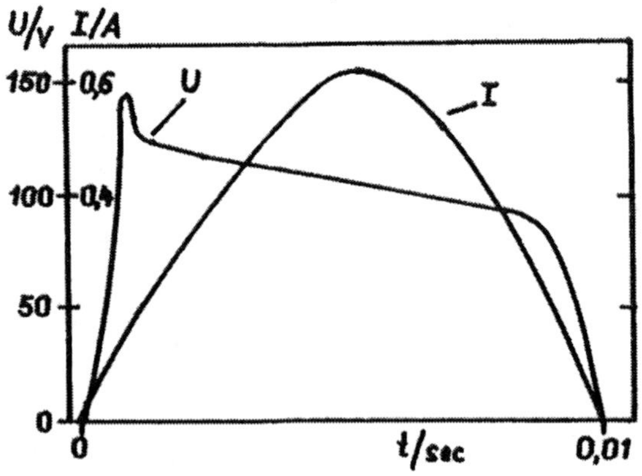

Fig. 2

References:

/1/ Polman, J.; J.E. van der Werf; P.C. Drop; J.Phys. D: Appl. Phys., 5 (1972) 266

/2/ Holstein, T.; Phys. Rev. 72 (1947) 1212; 83 (1951) 1159

THE ENERGY BALANCE OF THE 10 MM DIAMETER LOW PRESSURE RARE GAS-MERCURY DISCHARGE COLUMN

Shaolong Zhu, Beiyuan Zhu and Zuquan Cai
Institute for Electric Light Sorces, Fudan University.
Shanghai, China.

Because of the development of the compact fluorescent lamps, such as PL and SL lamps, the studies of small diameter low pressure rare gas-mercury discharge have risen great interest /1/. The column of such discharges was intensively studied: quartz discharge tubes were employed for measuring the absolute intensities of 253.7 nm UV radiation at various rare gas mixtures, bath temperatures and discharge currents, using deuterium lamp as a standard. The visible triplet was also measured using tungsten strip lamp as a standard. The electron temperatures were measured with Langmuir probes and the axial electric fields by floating probes. It was found that the efficiency of the useful radiation, i.e. the sum of the 253.7 nm and the visible radiations was a little higher than that of the 36 mm diameter discharges in certain discharge conditions. It was also found that the addition of small amount of krypton or neon may reduce or increase the axial electric field without measurable decrease of the useful radiation efficiency.

With a few modifications, Waymouth's analytical method /2/was used to calculate the energy balance of the small diameter discharge column. Because of the relatively large contribution of 185.0 nm radiation at high electron temperature in small diameter tube, its calculation should be refined: its imprisonment should be taken into account and the constant excitation cross section from 6 $^1S_0$ to 6 $^1P_1$ should be replaced by recently published excitation functions /3/ . For pratical applications the mercury vapour pressure was so high that its influence on electron mobilities must be included. The model was also extended to treatment of mixture rare gas-mercury discharges. All the cross sections used in the calculations were corrected by new published data. The results showed that the calculations of the column energy balance were in reasonable agreement with experiments.

It is concluded: with correct design, the useful radiation efficiency of small diameter rare gas-mercury discharge may be a little higher than that of conventional, and the addition of small amount of Kr or Ne to Ar-Hg discharge may effectively reduce or increase the axial electric field, also the lamp wattage without loss of useful radiation efficiency.

Reference

/1/. A. Bouwknegt, J.IES Vol.11 (1982) 204.

/2/. J.F.Waymouth, Electric Discharge Lamps, MIT press, Cambridge, (1971).

/3/. J.G.McConnell, B.L.Moiseivitsch, J.Phys. B, Vol.1 (1968)406.

THE INFLUENCE OF INERT GAS MIXTURES ON THE PROPERTIES OF COMPACT FLUORESCENT LAMPS

W.Pabst , OSRAM GmbH , D-8000 München 90

Argon is the most commonly used inert gas in fluorescent lamps . The use of other species such as Ne , Kr , He or mixtures of them can be an important tool in the modelling e.g. of electrical lamp properties for given geometrical parameters and constraints in the supply voltage . A well-known example is the lowering of lamp voltage in T8 lamps by the addition of Kr .

Detailed studies of the influence of inert gas mixtures have been performed on compact fluorescent lamps with T4- (12 mm-) diameter (type DULUX) . Fig.1 shows lamp voltage , lumen output , and efficacy variation with inert gas mixture Ne/Ar , Ar/Kr , and Kr/Ne . The lamps were operated with reference choke (type 7/9/11 W) on constant supply voltage (220 V 50 Hz ~) in an integrating sphere . Two series of measurements were performed with 4 lamps for each mixture , the total pressures being 4.7 and 3.3 mbar , respectively . As compared to Ar , Kr can reduce lamp voltage by as much as 28 % , whereas Ne can cause a voltage rise by as much as 31 % . Lamp power and lumen output change accordingly ; lamp efficacy reaches a maximum - besides for 100 % Ar - for Ne/Kr mixtures of 80/20 . The coldspot temperature not being held constant varied with lamp power up to about 7 degrees - in the case of constant temperature the efficacy curve would be slightly modified according to maximum output at about 45 °C for Ar 100 % .

Variation of total pressure beyond 3.3 and 4.7 mbar shows quite a different behavior of the pure gases Ne , Ar , and Kr . Lamp voltage rises with rising Ne pressure , but falls with rising Kr pressure . For Ar , a flat voltage minimum at about 6 mbar is found . Maximum efficacy is found - besides for 4 mbar Ar pressure - for low Kr and high Ne pressures . Besides the properties of the different gases with respect to electron and Hg collision and heat transport , the relative electrode losses play a certain role in efficacy considerations .

Admixtures of He cause a pronounced rise of lamp voltage - up to a factor of two - and thus of power density . This is an important fact for the "compactness" of a lamp . It is due to the absence of the Ramsauer effect in the elastic scattering cross-section of electrons for He ; in turn it causes reduced electron temperature and thus reduced lamp efficacy . Nevertheless , a good compromise is found for Ar/He mixtures of 50/50 at 2.4 mbar total pressure : as compared to standard Ar filling , up to 50 % higher lamp voltage , up to 30 % higher lumen output , and 5 to 20 % lower efficacy are found depending on lamp power . In the case of low pressure pure He filling , lamp life will be affected due to emitter material evaporation .

Experiments showed that He loss will not be critical at the wall temperatures of low pressure compact lamps .

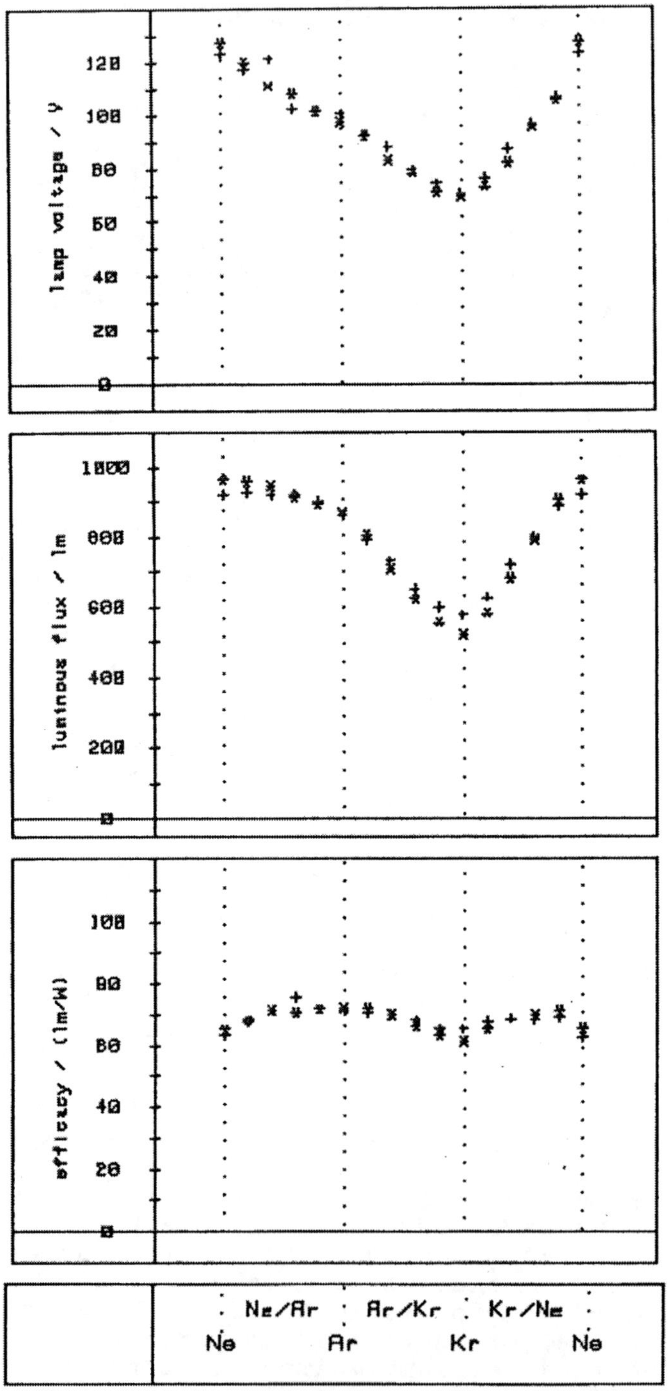

Fig.1. Lamp voltage , luminous flux , and lamp efficacy for various inert gas mixtures ; total pressure 4.7 mbar (*) and 3.3 mbar (+)

NR. 29

EXPERIMENTAL STUDIES OF THE ENHANCEMENT OF THE UV EMITTANCE OF A LOW DENSITY MERCURY DISCHARGE BY A MAGNETIC FIELD AND ISOTOPE BLENDING

Tai-Ming Zhou*, Fu-Zi Li*,
D.D. Hollister and S.M. Berman

Lighting Systems Research Group
Applied Science Division
Lawrence Berkeley Laboratory
University of California
Berkeley, California, U.S.A. 94720

The effects of applied magnetic fields and isotope enrhichment on the radiant emittance of the Hg-Ar discharge were described theoretically by Richardson and Berman(1). Subsequent experimental activities in this laboratory were inteded to aid in the application of this theory to the enhancement of the efficacy of fluorescent lamps by developing novel methods to make the resonance transition between the $^3P_1$ level and ground more efficient. Two such experiments and the results are described in this paper.

In the magnetic effects program, a uniform magnetic field is applied axially to the entire length of a temperature-controlled, low-pressure mercury arc that is rigidly mounted in an integrating cylindrical photometer. The total emission at λ2537 is observed both photoelectrically and spectroscopically as a function of cold spot temperature and applied magnetic field. It is found that the maximum enhancement in the production efficiency of radiation at λ2537 is 4.5%, with an applied axial magnetic field of 320 gauss and the mercury cold spot equilibrated at 40°C. Earlier observations(2) of UV emittance were made directly through a slit along a path at right angles to the applied magnetic field. With a temperature of 40°C, and with an applied field of the same order

This work was supported by the Assistant Secretary for Conservation and Renewable Energy, Office of Buildings Community Systems, Building Equipment Division of the U.S. Department of Energy under Contract No. DE-AC03-76SF00098.

*Permanent Address: Fudan University, Institute for Electric Light Sources, Shanghai, China

as above, the observed λ2537 production ∿20%. Spectrographic evidence is presented to show this discrepancy to be due to the angular dependence of the Zeeman shifted β components of the resonance line, and the depolarizing effect of the integrating cylinder photometer.

In the mercury isotope blending activity, the present effort is an examination of the enhancement of the efficiency of resonance radiation production as a function of the percentage enrichment of the natural mercury discharge vapor by the isotope ^{201}Hg. This activity is complicated by the necessarily small incremental amounts of added isotope and the moderately rapid mercury sorption rate by clean quartz, coupled with a need to know the relative ^{201}Hg concentration at all times. The experimental procedure is reviewed and current radiation emittance versus percentage ^{201}Hg content is presented and compared with similar measurements in which enrichment with ^{196}Hg is examined.

References:

(1) R.W. Richardson and S.M. Berman, "Theory of the Effects of Magnetic Fields and Isotope Enrichment on the Radiant Emittance of the Hg-Ar Discharge", presented at the Third International Symposium on the Science and Technology of Light Sources, Paul Sabatier University, Toulouse, France 18-21 April (1983).

(2) D. Hollister and S.M. Berman, "Experimental Studies of Enhanced Emittance by Zeeman Splitting of a Low-Pressure Mercury/Argon Discharge", presented at the Third International Symposium on the Science and Technology of Light Sources, Paul Sabatier University, Toulouse, France, 18-21 April (1983).

NR. 30

A SURFACE WAVE LAMP

D. Levy, S. Berman and D. Crawford

Lighting Systems Research Group
Applied Science Division
Lawrence Berkeley Laboratory
University of California
Berkeley, California, U.S.A. 94720

A low-pressure discharge was initiated and sustained by surface waves. The discharge container was a conventional fluorescent lamp discharge tube containing a mix of argon and mercury vapor. The range of r.f. frequencies used was 150-1000 MHz. Considerable improvement in lamp efficacy was observed.

In a low-pressure argon-mercury discharge tube with telectrodes at each end (such as a standard fluorescent lamp discharge tube), the electric power delivered per electron is greatest at the discharge tube's center. By way of contrast, in a plasma column sustained by surface waves, the electric power delivered per electron is greatest near the discharges tube's inner surface. Thus, the production of radiation (in particular, 253.7 nm) is enhanced near the outer edges of a plasma column sustained by surface waves, as compared to a standard positive plasma column. For an argon-mercury discharge, this implies that self-absorption of 253.7 nm radiation by mercury can be reduced using surface waves because the 253.7 nm radiation traverses a shorter path to reach the edge of the discharge tube as compared to a standard positive column.

The light output from an argon-mercury discharge tube surrounded by an r.f. shield whose inner surface was coated with cool white phosphor was compared when electric power was delivered to the plasma using surface waves, and when electric power was delivered to the plasma using electrodes and a standard ballast. In both cases, when identical amounts of electric power were delivered to the plasma in the discharge tube, the light output in the surface

---

This work was supported by the Assistant Secretary for Conservation and Renewable Energy, Office of Buildings Community Systems, Building Equipment Division of the U.S. Department of Energy under Contract No. DE-AC03-76SF00098.

wave mode was more than 30% greater than the light output when the discharge tube was powered in the conventional manner.

A unique energizer was designed to couple r.f. power to the discharge tube. This energizer provides instant starting upon application of r.f. power without a starting circuit. Subsequent to starting, it delivers 100% of the r.f. power (with no reflected power) in the form of cylindrically symmetric surface waves. Combining this energizer with an argon-mercury discharge tube and a transparent (either to visible or ultraviolet radiation) r.f. shield provides a surface wave lamp which requires no ballast, starting circuit, or electrodes.

ON THE ELECTRIC CHARACTERISTICS OF Ar + Hg DISCHARGE LAMPS WITH SUPER-NARROW DIAMETER

Z.L.Mi
Shanghai No.2 Special Lamp Works, Shanghai, China
C.H.Huang, D.Y.Fang, Z.Q.Cai
Electric Light Sources Institute, Fudan Univ., China

Super-narrow diameter Ar + Hg discharge lamps, because of their smaller tube diameters(~3 mm) and dimensions, have been widely used for special purposes, such as in the drilling photography and modern physics experiment(Zeeman effect experiment). The subject of this paper is to investigate the electric characteristics of the Ar + Hg discharge lamp with super-narrow diameter. Fig.1 is the sketch of the lamp.

Fig.1 The Ar + Hg discharge lamp with super-narrow diameter.

1. Breakdown characteristics

It is found that the dependence of the breakdown voltage V_s on the product of the argon pressure and the electode spacing pd obeys qualitatively Pachen's Law, though the tube radius is very small. In order to get lower ignition voltage, we find that the argon pressure, for a 3.2 mm diameter lamps, should be limited in a region

$$7.0 \text{ Torr} \leqslant P_{Ar} \leqslant 13.0 \text{ Torr}$$

2. V-I characteristics

Fig.2 presents the typical voltage-current characteristics of a super-narrow diameter Ar + Hg discharge lamp connected in series with a choke in a 50 Hz supply, the tube wall of the discharge lamp was controlled by a thermostat.

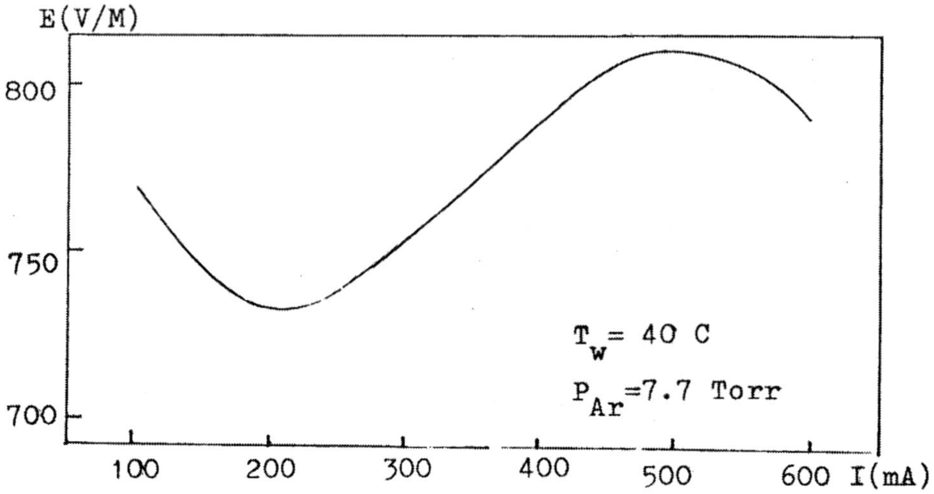

Fig.2 The electric field strength vs electric current of a super-narrow diameter Ar + Hg discharge lamp.

From this figure, it can be seen that the electric field strength increases when the discharge current is beyond a certain value; when the discharge current increases further, the electric field strength begins to fall down. It appears that when the discharge current is larger than a certain value, the E-I characteristics of the Ar + Hg discharge is similar to that of low pressure sodium discharge when the discharge current reaches the critical value. Therefore, the measured E-I curve here for the Ar + Hg discharge seems to aries from the mercury deplition at a current density higher than 3×10^4 A/M^2.

ON THE OPTIMUM COLD SPOT TEMPERATURE OF THE FLUORESCENT DISCHARGE LAMP

D.Y.Fang, C.H.Huang, Z.Q.Cai
Electric Light Sources Institute
Fudan Univ., Shanghai, China

For an optimum design of fluorescent lamps it is important to know the dependences of the optimum cold spot temperature (or the optimum mercury density $N_o(opt)$) on the tube radius R and the buffer gas pressure P_{Ar} of the discharge lamp.

From Eq.(27) of recently developed Lama's model[1] on Ar + Hg discharge, let $\partial P/\partial N_o = 0$, one can obtain the expression for the optimum mercury density as follows

$$N_o(opt) = \frac{1}{C_3 \cdot R \cdot N_e(P_{Ar}, I)} \qquad (1)$$

where $C_3 = 9.9 \times 10^{-37} m^5$, R is the tube radius (mm) and N_e is the electron density (m^{-3}) which is dependent on the pressure P_{Ar} and the electric current I of the discharge lamp.

From Eq.(1) it can be seen that the $N_o(opt)$ is inversely proportional to R and $N_e(P_{Ar}, I)$ thus P_{Ar} (at constant tube radius and electric current, with decreasing P_{Ar}, the electron mobility will increase, therefore, N_e will decrease, from Eq.(1), $N_o(opt)$ will increase).

Fig.1 shows the measured dependences (dots) of the $N_o(opt)$ on N_e, R and P_{Ar}. It can be seen that the theoretical curves (dashed lines) calculated according to Eq.(1) agree qualitatively with the experimentally measured one.

Unfortunately, quantitatively there is a great deviation between the experimental results and the calculated one, especially at lower electron density, smaller tube radius and lower buffer gas pressure, which arises from the simple hypotheses in Lama's model.

By refining Lama's model, we have

$$N_o(opt) = \frac{1}{C_3 \cdot A(N_e,R) \cdot R \cdot N_e(P_{Ar},I)} \quad (2)$$

where $A(N_e,R)$ is a correcting factor obtained by fitting the experimental data, it is given by

$$A(N_e,R) = \left(2.310 - \frac{35.313(N_e' - 1)}{15.860 \times N_e' - 1}\right)\left(\frac{18 - R}{1.227R + 1.948} + 1.192\right)$$

where $N_e' = N_e \times 10^{-17} m^{-3}$.

Fig.1 shows also the calculated results (solid lines) according to Eq.(2). One can see that the present calculated results coincide with the experimental data much better than that of Lama's model. In addition, by substituting $C_3A(N_e,R)$ for C_3 in Eq.(27) of Lama's model, the calculated dependence of the 2537 Å radiation output on the cold spot temperature agrees with the experimental data better than that of Lama's model, as illustrated in Fig.2.

Ref.1. W.L.Lama, et al., Appl. Opt. 21 1801 (1982)

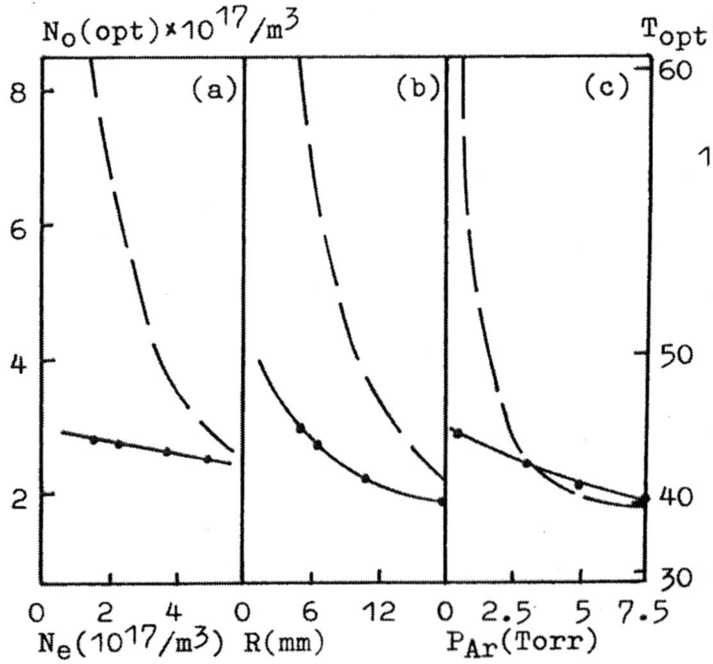

Fig.1 $N_o(opt)$ vs N_e, R and P_{Ar}

(a) R=7mm, P_{Ar}=3Torr

(b) P_{Ar}=3Torr, $N_e=2.5\times10^{17}m^{-3}$

(c) R=7mm, $N_e=2.5\times10^{17}m^{-3}$

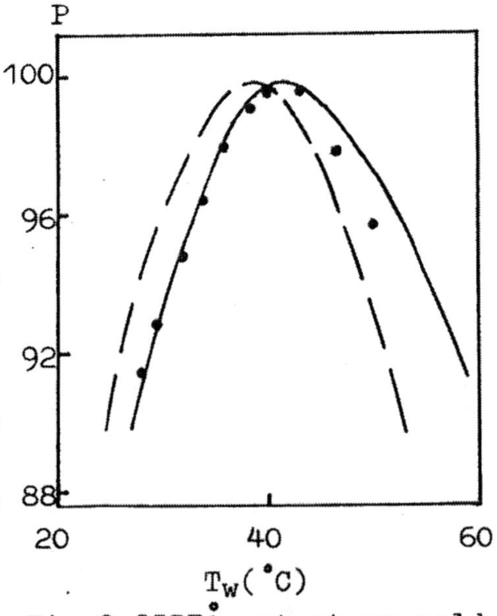

Fig.2 2537Å output vs cold spot temperature

THE RUN-UP CHARACTERISTICS OF VARIOUS COMPACT FLUORESCENT LAMPS

W. Hellebrekers and J. Heuvelmans
Nederlandse Philips Bedrijven B.V.
Lighting Division
Eindhoven, Netherlands

When considering the run-up behaviour of fluorescent lamps they can be divided into three main classes. First is the well known linear or circular fluorescent lamp in which the mercury vapour pressure is controlled by the wall temperature. The second is a more compact fluorescent lamp with one or more so called cold spot(s) (e.g. 'PL', 'DULUX') which controls the mercury vapour pressure. The third type is a compact fluorescent lamp in which the mercury vapour pressure is controlled by an amalgam (e.g. 'SL'). A short review is given of the run-up characteristics of these three classes of lamps.

1. The run-up of a fluorescent lamp after ignition is determined by the increase of the mercury pressure from room temperature to its stable value. The mercury vapour pressure depends on the wall temperature of the discharge vessel. This wall temperature again depends on the amount of power which is dissipated in the lamp, the diameter of the lamp and the ambient temperature. In general the lamp is designed so that at stable operation the wall temperature corresponds with the optimum mercury vapour pressure. Knowing the relative luminous flux as a function of the wall temperature and knowing the run-up of the wall temperature, the run-up of the luminous flux of the lamp can be

predicted very well. In order to eliminate electrical interaction with the control gear such experiments are carried out with a constant electric current.

2. Compact fluorescent lamps (e.g. 'PL', 'DULUX') often use a so called cold spot in order to stabilize the mercury vapour pressure at the optimum value during stationary operation. All the mercury in the lamp will be collected at this cold spot. The run-up of this lamp is more complicated. Experiments have shown that the run-up also depends on the off-time of the lamp, the type of phosphor coating and the temperature distribution of the whole discharge vessel.
3. The mercury vapour pressure in a compact fluorescent lamp can also be controlled by the use of an amalgam (e.g. 'SL'). In this way the luminous flux can be kept close to its maximum value over a large temperature range. Generally this temperature range is about 2 times larger than for a pure mercury lamp. In addition to a (main) amalgam it is necessary to use an auxiliary amalgam. This auxiliary amalgam improves the initial run-up of the lamp. The type and amount of the auxiliary amalgam and the main amalgam and its temperatures will determine the run-up characteristic of the lamp. Also the off-time of the lamp is important.

For all these situations the results of well controlled experiments will be presented and the associated dynamic behaviour of the mercury vapour pressure will be discussed.

EFFECT OF END BANDS ON LUMEN MAINTENANCE OF FLUORESCENT LAMPS

E.J.Covington

Lighting Systems Products Division
General Electric Company
Nela Park
E. Cleveland, Ohio 44112
U.S.A.

It is usually assumed that the darkening that occurs at the end of a lamp does not subtract significantly from the integrated light output. This assumption is generally true but in some cases the visual defect can be significant and should be considered in electrode design and lamp processing.

The appearance of the lamp end discoloration is determined to a great extent by the carbonate decomposition schedule used during manufacture. When higher-than-normal electrical currents are used the end band is the most prominent type of discoloration observed. When lower-than-normal currents are used the spot type of discoloration in the immediate area of the electrode is observed. The end band usually has a sharp boundary in the transition region between the Faraday dark space and the positive column. The intensity of the darkening decreases as the center of the lamp is approached, however this type of darkening can result in a significant reduction in total light output as well as an objectionable visual appearance.

A mechanistic description of the formation of these bands(or rings) is given. This usually entails the generation of oxygen in the emission mix and then the formation of HgO in the region in front of the electrode. The contaminants are considered from the standpoint of their composition (HgO+Hg) and light absorption characteristics. Other processes that can occur during lamp operation, such as evaporation, sputtering, and dissociation of oxides by electrons and ions, are considered for possible contribution to the formation of the end bands.

The types of discoloration that are located in the vicinity of the electrode usually do not have a strong effect on the integrated light output because even in the absence of darkening the end of the lamp does not have a high output. This condition exists because it requires a tube length that is equivalent to about seven tube diameters before the light level is at its maximum value; this is the result of internal reflections. On the other hand, end bands are located at the start of the positive column where the light level is closer to the maximum value. Calculations show that a contaminate film of about ten monolayers thickness is able to absorb about 20% of the impinging 254nm radiation.

Lamp end-to-end light profiles were taken as a function of time, decomposition schedule, and type. The types used for test vehicles were the 420 ma four-foot 1.5-inch diameter linear lamp, as well as a 300 ma 8-inch diameter circular lamp with a bulb diameter of 1.12 inches.

End bands that are noticeably visible on a four-foot 1.5-inch diameter lamp can reduce the integrated light output by 5-6%. Light reduction due to the formation of end bands occurs before the darkening is visually perceptible. The band becomes perceptible when the local light level is diminished by about 10% in value. The effect of end bands on lumen maintenance increases, of course, with a decrease in the length-to-diameter ratio of the glass tube. It should be mentioned that the main reason for the decrease in light output from a lamp is due to the effects of mercury ion and UV radiation impingement on the phosphor surface.

THE USE OF MULTIPLE ELECTRODES TO OFFSET THE GLOW TO ARC TRANSITION IN LOW PRESSURE DISCHARGE LAMPS

J.E. HARRY

Department of Electronic and Electrical Engineering

University of Technology, Loughborough, Leicestershire, U.K.

Conventional discharge lamps use a single pair of electrodes. For a single pair of electrodes the maximum discharge current is limited by the tendency for the glow to arc transition to occur and cause collapse of the glow discharge. As a result an upper limit exists for the current input to glow discharges.

The critical region in suppressing the glow to arc transition is usually at the anode root at which the current density is normally higher than at the cathode. The current at which the discharge degenerates into an arc increases as the number of anodes is increased. The use of multiple anodes is made possible by connecting separate stabilising impedance in each anode supply circuit. A large number of discharges sufficiently separated so that they do not coalesce with individual anodes and a common cathode has enabled the total current of a glow discharge to be increased to several amps.[1]

The current density at the cathode usually remains constant provided the area of the cathode is adequate. The current density increases with current when the cathode surface is covered. If the current density at the cathode exceeds a critical value depending on the operating conditions then the discharge will collapse in the region of the cathode fall and degenerate to an arc. The glow to arc transition may be also initiated by constriction of the discharge in the cathode fall region or the onset of filamentary discharges in the discharge column resulting from an excessive voltage gradient to neutral particle density ratio $(\frac{E}{n})$ or non-uniform current distribution within the column.

Increasing the number of separately stabilised cathodes distributes the current equally between the cathodes and enables the discharge to be operated at a higher total current. The use of multiple cathodes also results in a reduction in the fall voltage in the abnormal glow region and a decrease in sputtering of the cathode material, an important factor in determing the life of some discharge lamps, where the effect of sputtering is to poison the gas and coat the lamp walls.

Any number of parallel discharges can be operated stably in close proximity to that they coalesce and interact electromagnetically. This results in the production of a large volume of ionised gas no longer limited by the diameter of a single discharge.[2] The electrical circuit is shown in Fig.1. The electrodes

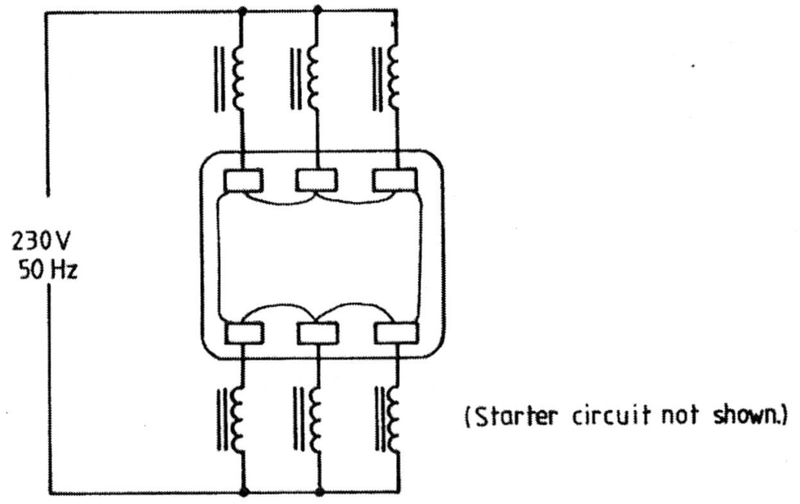

Fig.1. Multiple discharge lamp and its stabilising circuit.

are individually stabilised with inductive ballasts of half the normal impedance and the current is shared equally between them. A low pressure lamp with six thermionic electrodes has been operated in this way under the usual operating conditions of a glow discharge lamp but with three times the normal operating current.

REFERENCES

1. Harry, J.E., Knight, R., Multiple electric discharges, Proc. IEE, Jan. 1985.
2. Harry, J.E., Electric discharge lamps with multiple electrodes, 3rd International Symposium on the Science and Technology of Light Sources, 18 - 21 April, Toulouse, France, pp. 86-87, 1983.

IMPROVED VACUUM IN LOW-PRESSURE SODIUM LAMPS

B.J. de Maagt and J.J.G. Willems
Nederlandse Philips Bedrijven B.V.
Lighting Division, Eindhoven, The Netherlands

In the evacuated outer bulb of low-pressure sodium (LPS) lamps a barium flash getter is used in order to remove residual gases such as hydrogen, nitrogen, oxygen, water vapour, carbon monoxide and carbon dioxide. Noble gases are not captured. Surprisingly, also methane, which appears in considerable amounts after evaporation of the barium getter, is not absorbed.

Methane is formed by the reaction of hydrogen or water vapour with traces of carbon on the barium mirror. Avoiding the formation of methane or eliminating methane will improve the heat insulation of the discharge tube and as a result the lamp power comsumption will be reduced, while the luminous efficacy increases.

It has been found that methane formation cannot be avoided by replacing the barium getter by a titanium mirror (1). Because of the complexity of the electrical circuit thermal cracking of methane by means of an auxiliary tungsten filament in the outer bulb (2), is not a very practical solution. Therefore, a search was made for a chemical method to remove the methane.

We found that several metals and metal oxides present in a highly dispersed state on a porous support of alumina or silica catalyze the decomposition of methane at moderate temperatures (3). The best results are obtained with a commercial 0.5 % Pt on alumina pellet.

Figure 1 shows the methane pressure evolution during the first stage of operation for two SOX-E 66 lamps, one with and one without a methane decomposition catalyst. The catalyst pellet (40 mg; ø = 1/8 inch) was mounted near the U-bend of the discharge tube, where a temperature of about 200°C will be reached. There were no additional modifications to the manufacturing process. In the lamp without the catalyst the methane pressure increases continuously.

On the other hand, the pressure in the lamp with the decomposition catalyst initially increases, but subsequently decreases to less than 10^{-4} mbar within 15 hours.
As a result of this improved vacuum the power consumption of the LPS lamps SOX-E 66 and SOX-E 131 is reduced with about 10 W.

Mass spectrometric analyses of lamps equipped with such a catalyst showed that after 10,000 hours in operation the gas mixture, with a pressure $p \leq 2.10^{-5}$ mbar, only contained argon and helium.

The methane decomposition catalyst thus proved to be effective throughout the whole lamp life.

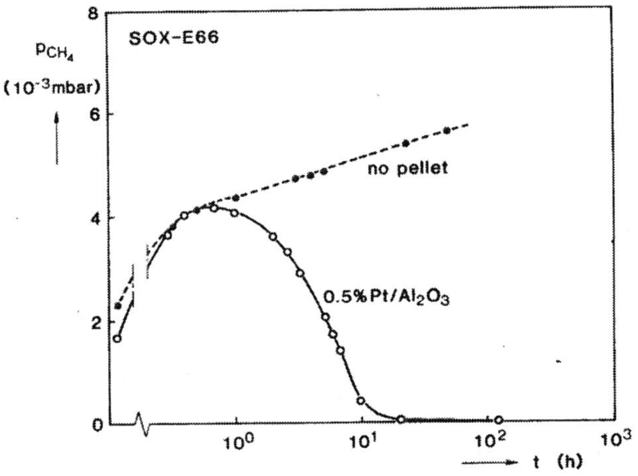

Fig. 1: Methane pressure evolution in outer bulbs of SOX-E 66 LPS lamps with and without a methane decomposition catalyst.

REFERENCES

1. W. Kroontje, B.J. de Maagt, M.H.J. van Rijswick and L. Sprengers, to be published.
2. Dutch Patent Appl. 7907220.
3. Dutch Patent Appl. 8201750.

INCANDESCENT LAMPS

37 - 45

MODEL FOR DESIGN OF INCANDESCENT LAMPS

Rolf S. Bergman
General Electric, Lighting Business Group
Nela Park, Cleveland, Ohio, 44112

A predictive design model for incandescent lamps, based on fundamental physical principles, has been developed and coded for computer use. The computer code can be used to develop filament designs based on prescribed performance requirements, and/or predict changes in performance of given filament designs due to changes in variables such as voltage or gas type and pressure.

The model structure assumes that seven independent variables, i.e., voltage (current), power (resistance), lumens (efficacy), life, wire diameter, wire length, and filament temperature, must be defined to obtain a given lamp design. However, only four independent equations, all functions of wire length, wire diameter, and filament temperature, can be identified. These are equations for a) wire resistance, b) filament energy (power) balance, c) lumen output, and d) life. Since the number of variables exceeds the number of equations, three of the variables must be specified in order to obtain a unique solution. In addition, a number of other parameters, such as coiling, gas type and pressure, envelope geometry and temperature also must be specified for each design.

One of the features of the present model that enhances its accuracy as a predictive tool over a wide range variables is that the energy balance equation accounts for every major loss process. The radiative losses are obtained by spectral integration of Planck's law multiplied by the spectral wire emissivity and a coiling reabsorption factor. Gas convection losses are evaluated using the "Langmuir sheath" formulation, where the sheath edge is calculated using convective heat transfer methods. The conductive losses to the leads are determined by a simplified energy balance of the heat flow at the filament-lead junction.

Another important feature is the equation for filament life. It is derived by assuming that life is inversely proportional to the diffusive flux of tungsten from the coil and also dependent on the wire diameter in an empirically determined manner. The equation is normalized by assuming that the 100W household lamp life 750 hours.

The variables that satisfy all four equations for the given input constraints are obtained through iteration, the exact manner depending on which variables are initially specified. For example, if voltage, power, and life are given, the wire diameter and length are calculated by iteration of the resistance and energy balance equations for a fixed temperature. The temperature is then recalculated using the life equation and the process repeated until convergence is obtained.

Computed results, for various design specifications, including high pressure incandescent lamps, will be presented and compared to measurements. In conclusion, extensions of various aspects of the model and the type of future work required to strengthen and verify the model for incandescent (and halogen) lamps will be discussed.

Aging Studies of Low Voltage Incandescent Lamps for Long Duration Space Applications

H. W. Leidecker, C. E. Powers, and W. Viehmann
NASA Goddard Space Flight Center
Greenbelt, MD, 20771, U.S.A.

Aging characteristics and lifetimes of 1.5 watt krypton-filled lamps with straight singly coiled filaments of 135 μm inside diameter wound at 16 turns per mm from 38 μm diameter wires of tungsten, tungsten+3%rhenium, and tungsten+3%thorium have been evaluated as functions of voltage and filament length. For several years, lamp voltages, currents, and the light flux in two directions (measured by Si photovoltaic cells) have been automatically recorded daily with computer controlled equipment with precisions better than 0.1%. High spatial resolution radiometer filament scans have been taken periodically to detect and to follow the development of "hot spots", "cold spots" (turn-to-turn shorts), and filament deformation. High resolution SEM micrographs and EDAX scans were taken after burn-out.

The voltage dependences of the current, the light fluxes, and the center filament temperature have been measured for most lamps after a one-month burn-in, and empirically found to be power laws over the range from 3 to 6 volts (1900 to 2500 °K center filament temperature): Property = Constant * $\text{Voltage}^{\text{Power}}$. The fitting parameters for the current-voltage relation have been found to be a sensitive indicator of the filament material.

All lifetime studies have been conducted at constant voltage, at filament temperatures between 2000 and 2400 °K. By the end of the burn-in, the central 80% of the filament is a single crystal, which joins poorly with the colder polycrystalline ends. Notches grow into the filament along certain crystallographic planes which repeat with four-fold symmetry along the filament. At first, notch growth is driven by a combination of the Soret effect and electromigration, and not by evaporation. The growth rate of these notches is uniform when they are shallow, and is a definite function of the electrical current, the temperature distribution, and the filament alloy. When a notch becomes deep enough, the local temperature becomes hot enough for evaporation to dominate, and the growth of that notch accelerates until either burn-out or re-welding occurs, with roughly equal chances. These events are clearly visible in the daily current data, and their rates have been measured. As many as ten re-weldings have been observed to occur in a single lamp. The on-set of evaporative acceleration is unpredictable in detail, as it depends on random irregularities in the filament. Hence, the lifetime of a lamp is a doubly random process: first, some notch must become deep enough for accelerative evaporation to occur; second, a separation and not a re-welding must occur. An empirical relation has been developed for tungsten filaments which relates average lifetime to filament voltage and filament length. This relation is in excellent agreement with lifetimes observed in the GOES series of satellites.

Nr. 39

THE DESIGN, CONSTRUCTION AND PERFORMANCE OF AN INCANDESCENT LIGHT SOURCE WITH A TRANSPARENT HEAT MIRROR

R. P. Howson, R. P. Fontana, I. S. Goldstein and L. Thorington

Durotest Corporation

North Bergen

New Jersey 07047, U. S. A.

The factors in the design of an economical incandescent light source using a filter envelope which directs heat back to the filament but allows light to be transmitted are discussed. The achievement of the necessary performance requires a new concept in the design of the lamp and the achievement of a high performance coating. The use of reactive sputtering to prepare dielectric-silver-dielectric filters in special designs is shown to give lamps with twice the luminous efficiency of a standard unit.

BOUNDARY CONDITIONS FOR REGENERATIVE CYCLES IN HALOGEN-FILLED INCANDESCENT LAMPS

G. Dittmer

Philips GmbH Forschungslaboratorium Aachen, Germany

Tungsten and molybdenum transport occurring in halogen-filled incandescent lamps may be calculated by means of a computer program solving chemical equilibrium equations. The procedure is based on minimization of the free enthalpy of the system

$$G = \sum_{j=1}^{n} \mu_j N_j$$

regarding relevant mass balance equations

$$\sum_j \vec{\nabla} \nu_{ji} \frac{P_j}{RT} \left[\vec{\nabla} P_j + P_j \alpha_j \frac{\nabla T}{T} \right] = 0.$$

The transport properties are influenced by the effective gas phase composition which is determined by the initial gas filling, the presence of solid compounds and the temperature distribution within the lamp.

The presence of solid tungsten or molybdenum at low temperatures may lead to the formation of solid phases of the general composition MO_2, MO_3, MOX_2, MOX_3, MOX_4, and MO_2X_2 and after a prolonged period of lamp operation to compounds such as MX_2, MX_3, and MX_4. Generally the formation of solid phases is enhanced by lowering the bulb temperature and by increasing oxygen and halogen pressures and counteracted by an excess of hydrogen.

A regenerative cycle in lamps only can be maintained if the temperatures of the bulb or the current leads allow to retain sufficient amounts of halogen in the gas phase. Furthermore an excess of hydrogen leading to a reduction of of transport-supporting tungsten compounds has to be avoided. In fig. 1 the lowest admitted temperature for a regenerative cycle is given as a function of filling pressures.

In most halogen-filled incandescent lamps a competition between tungsten and molybdenum transport occurs. As demonstrated in fig. 2 the tungsten transport cycle is not noticeably influenced by molybdenum at lower temperatures; at higher temperatures, however, the molybdenum transport prevails. Consequently the temperature of molybdenum reflectors and current leads should be kept below 1400 K.

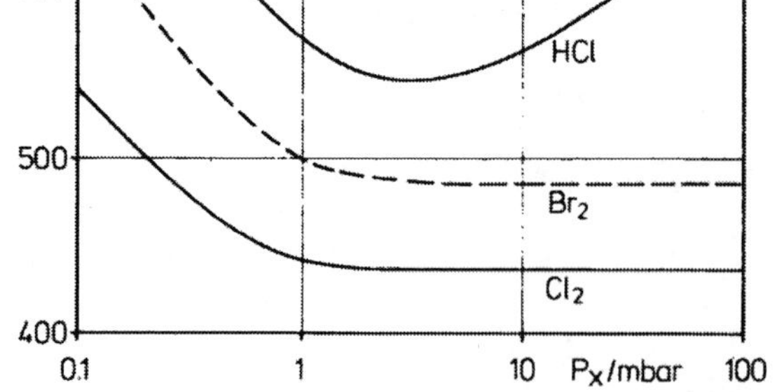

Fig. 1: Lowest bulb temperatures for a regenerative cycle as a function of initial halogen or hydrogen halide pressures (with 0.1 mbar O_2, 5 bar Kr)

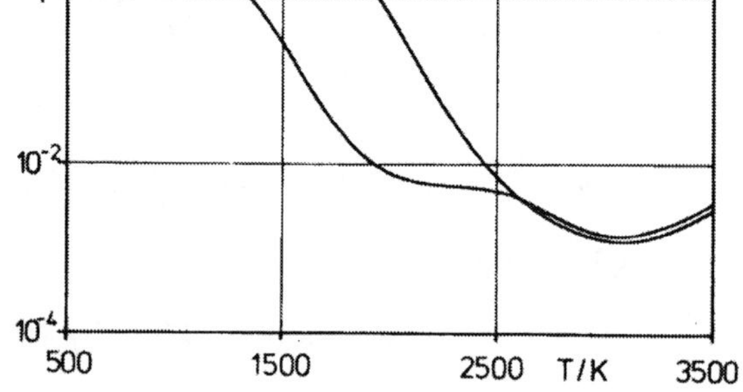

Fig. 2: Ratio of summed pressures of tungsten and molybdenum as a function of temperature (with 0.1 mbar O_2, 5 bar Kr)

SOLID COMPOUNDS IN HALOGEN-FILLED INCANDESCENT LAMPS

U. Niemann

Philips GmbH Forschungslaboratorium Aachen, Germany

The effective gas phase composition in halogen-filled incandescent lamps is significantly influenced by condensing solid phases. This holds especially for lamps with bulb wall temperatures below 600 K. In order to acquire a detailed knowledge of the composition and chemical behaviour of these phases transport experiments have been carried out under quasistationary conditions: Tungsten coils were mounted in large glass bulbs before filling with reactive gases (Cl_2, Br_2, Cl_2/O_2, Br_2/O_2). By varying filament and wall temperatures a number of solid halides and oxyhalides of tungsten and molybdenum could be produced during and after lamp operation.

In the W/Cl_2 system, for instance, at lower filament temperatures ($T_f \sim 1500$ K) WCl_6 precipitates at low bulb wall temperatures ($T_b \sim 300$ K), whereas at higher bulb temperatures ($T_b \sim 700$ K) the lower-valent halides WCl_4 and WCl_2 are formed. At higher filament temperatures ($T_f \sim 3000$ K) low-volatile WCl_2 deposits at the current leads at low and high bulb temperatures; additionally some WCl_6 condenses on the bulb wall after lamp operation.

In the $W/Cl_2/O_2$ system at $T_f \sim 1500$ K and $T_b \sim 300$ K the oxyhalides $WOCl_4$ and WO_2Cl_2 are formed; at $T_b \sim 700$ K precipitates of $WOCl_3$, $WOCl_2$ and $WOCl$ appear together with some volatile WO_2Cl_2. At $T_f \sim 3000$ K and $T_b \sim 300$ K WO_2Cl_2 condenses during lamp operation, whereas at $T_b \sim 700$ K WO_2Cl_2 and $WOCl_4$ precipitate only after lamp operation.

From these studies thermochemical data have been derived adequate for equilibrium calculations. For each type of lamp a specific level of summed tungsten in the gas phase is required for a regenerative cycle. Corresponding minimum temperatures for the bulb wall and the current leads in dependence from existing solid phases are given in fig. 1.

If molybdenum is introduced as a material for current leads the minimum temperature inside the lamp may be lowered by about 100 K: Due to the lower stability of solid molybdenum halides and oxyhalides increased amounts of halogens are present in the gas phase generating a summed tungsten pressure sufficient for a halogen cycle.

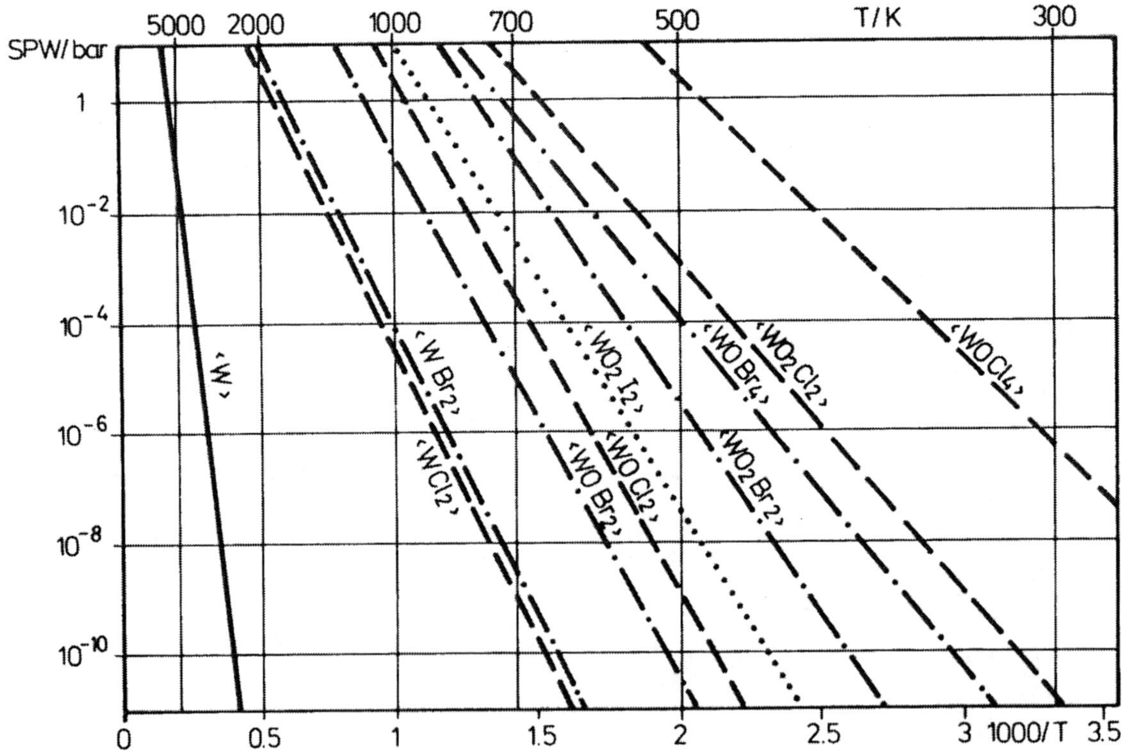

Fig. 1: Summed tungsten pressures above solid phases

EFFECTIVE MIXED HALOGEN DOSING OF LONG LIFE LINEAR TUNGSTEN HALOGEN LAMPS

I Connor and V W Goddard, THORN EMI Lighting Limited,
Research & Engineering Laboratories, Melton Road, Leicester, LE4 7PD, U.K.

Long life linear tungsten halogen lamps have traditionally been dosed with iodides in the knowledge that iodine does not cause tail erosion and, provided enough iodine is added in conjunction with oxygen in parts per million quantities, then the lamp will run clean to 2000 hrs. In practice most linear lamps become darkened with deposited tungsten well before the end of life.

In an attempt to reduce bulb wall darkening bromine is used as an alternative dose. However due to its increased activity it requires very good quality control of halogen, filament impurities and filling gas in order to obtain a satisfactory life. A partial solution to the high bromine activity is achieved by providing the filament tails with sacrificial overwind material. Previous work has also shown the possibility of using mixtures of iodine and bromine to control activity (1).

In this investigation the use of a buffered mixed halogen system has been assessed in 240V and 500W linear lamps, with the objective of obtaining the best features of both halogen systems. Our experiments showed that if iodine was added to bromine in a lamp, the activity of the bromine was modified. Hence this allowed excess bromine to be added without incurring bromine overactivity. By thus reducing the overall iodine content an increase in efficacy of 4-5% was achieved.

Iodine and bromine dosing was made via the introduction of iodoform and trimeric bromophosphonitrile dissolved in hydrocarbon solvents. Dose levels were optimised by preparing lamps with bromine compound levels between 25 and 150 µg and iodoform levels between 0 and 200 µg.

Severe tail erosion problems were experienced with lamps in the early part of this study. The erosion was not typical of that produced by overactivity, but occurred as very localised pitting of the filament tails. This type of erosion has previously been reported in literature (2), and a patent (3) describing the use of a tungsten/rhenium alloy as a solution has been granted. Our investigation showed that the pitting was associated with the presence of platinum on the tail surface. Examination of the in their formation using SEM/EDAX showed that the centre of the pit was rich in platinum. The presence of platinum resulted from its use as an intermediate in the formation of the weld between the filament and molybdenum foil. Changing to a tantalum intermediate or an intermediate free weld completely eliminated this type of tail attack.

After elimination of the tail erosion an optimum dose was based on the level of bromine to give a clean lamp at end of life together with the minimum iodine content to give sufficient buffering effect. This optimum dose required 75 µg bromine and 110 µg iodine, although concentrations either side of these figures gave satisfactory performance. It was established that a slight stoichiometric excess of bromine over iodine gave the best results.

JANAF data indicates that IBr has gas phase stability to temperatures in excess of those experienced at the filament tails in lamps with the equilibrium such that some bromine will always be available for reaction. It is believed that the halogens combine to form iodine monobromide which provides a buffering effect, thus enabling the use of bromine to eliminate through bulb blackening whilst retaining long life.

References

1. Magos L, paper read at Hungarian Technical and Scientific Days, 11-15 December 1967 (not published).

2. Rouweler, G, de Maagt, B J, Philips Journal Research, Vol. 33, 1978, 1-19.

3. UK Patent GB 2079531A.

The Use of Phosphorus as an Oxygen Getter and Halogen Cycle Modifier in Incandescent Tungsten-Halogen Lamps

Mark D. Beschle
GTE Lighting Products
Engineering Center, 60 Boston St., Salem, Mass. 01970

Recent theoretical and experimental investigations into the use of phosphorus as an oxygen getter in tungsten-halogen incandescent lamps has led to both substantial improvement in tungsten-halogen lamp life and to the development of an improved understanding of the thermodynamics of the tungsten-halogen cycle.

Although phosphorus has long been used as an oxygen getter in incandescent and tungsten-halogen lamps, the recent developments at GTE Lighting Products Engineering Center in Salem, Mass. have enabled the life and consistency of a wide range of incandescent tungsten-halogen lamp types to be substantially improved. In order to achieve these advances it has been found to be both theoretically and empirically necessary to supply phosphorus to the lamp in extremely precisely controlled amounts and purities.

Since traditional solid dosing techniques were found to be incapable of achieving the necessary control of these parameters, new processes were developed to meet our requirements.

The dose of phosphorus introduced into a given lamp type is specific to that lamp type and is generally specified and controlled within five(5) micrograms or better. The phosphorus utilized in the process contains impurities in the amount less than 1ppm.

A particularly useful tool in both developing a general understanding of the tungsten-halogen cycle and in predicting or specifying optimum phosphorus doses for specific lamp types has been a thermodynamic modeling program, CEC76. Utilizing this program it has been possible to suggest relative phosphorus dosages based on lamp operating temperatures and halogen dose. Use of this program has also enabled the prediction of lamp types which either are not benefitted by or are adversely affected by the use of phosphorus. Still another interesting example of both the usefulness of phosphorus and of the CEC76 program involves the interaction of phosphorus and carbon. In this case, utilizing information from the program it was predicted that the combination of phosphorus and carbon, in a halogen lamp, would result in a substantial reduction of the minimum operating temperature of the tungsten-halogen cycle. The experimental results establish the validity of this forecast. Finally, a combination of theoretical calculations from this model and empirical observations of over 3000 tungsten-halogen incandescent has forced a reconsideration of the role of phosphorus in the halogen cycle specifically, and more generally of the critical role of oxygen in the halogen cycle.

Phosphorus, if properly understood and utilized, can be an extremely benneficial additive to tungsten-halogen lamps both in extending the lives of a wide range of tungsten-halogen lamp types and as an investigative tool in furthering understanding of the thermodynamics of the tungsten-halogen cycle.

HALOGEN LAMPS BY HARDGLASS TECHNOLOGY

A. Bodmer, R. Kiesel and G. Maier
OSRAM GmbH
D-7922 Herbrechtingen

As the first lamp manufacturer OSRAM started in 1975 to develop a hardglass H4 halogen lamp for automated production lines. OSRAM-glass 943 proved to be suitable for this application due to its physical and chemical properties as well as to the technical and economic requirements. In the meantime other suitable types of glass are available: CORNING 1720 (limited) and 1724 as well as SCHOTT 8252.

The operation temperature of a H4 lamp can reach 640 °C. Therefore these glasses have strain points in the range of 660 to 680 °C. By coefficients of thermal expansion of 4.2 to 4.5 . 10^{-6}/K direct sealing of molybdenum wires in bulbs of these glasses is possible. The seal is of pinch type.

A very low sodium content is necessary not disturbing the tungsten halogen cycle. Progress is made to reduce sufficiently the contents below 0.15 %.

Also the tolerances of the tube dimensions as well as the optical quality equal those of the previous used "Vycor".

Now, this technology enables a direct and very reliable sealing of molybdenum wires on sealing machines at an index speed of faster than 1 s. A proper control of the glass temperature during the sealing is as important as a thermal stress-free annealing of the pinch.

OSRAM has also applied the hardglass technology to other halogen lamp types, such as headlight bulbs for the US market, and to a wide range of halogen MINIWATT lamps.

HALOGEN INCANDESCENT LAMPS WITH ELLIPSOIDAL REFLECTORS FOR THE ULTRAVIOLET AND BLUE SPECTRAL RANGES

K. Pescht
OSRAM GmbH
D-1000 Berlin 65

Low voltage halogen incandescent lamps with ellipsoidal reflectors (HER-lamps) are gaining in importance for equipment that is used for curing of polymeric materials, particularly in the dental medicine (plastic fillings) and increasingly for technical applications (polymerization glue).

Two new lamp types for the ultraviolet and blue spectral ranges are presented.

1. Introduction

The required radiation in the desired spectral range can be generated by HER-lamps. Most of the manufacturers today are using two methods in producing HER-lamps with radiation output in the ultraviolet or blue spectral ranges, resp.:

- Cold mirrors in combination with appropriate optical filters,
- selective reflecting dichroic mirrors.

The radiation is focused at flexible or rigid optical fibers that are brought into contact with the polymeric material for an exposure time of 15 to 30 s.

2. HER-lamps with ultraviolet or blue dichroic mirrors

HER-lamps with selective reflecting dichroic mirrors are more and more requested. Those lamps should fulfill the following requirements:

- Sufficient radiation output in the desired spectral range,
- tolerable decrease of radiant intensity during lamp life time,
- small radiation output in the infrared to avoid heat stress,
- stand-by operation,
- negligible amount of irradiation below 320 nm to aboid skin burn (erythema), particularly important with dental applications.

The above listed requirements are fulfilled by low voltage halogen incandescent lamps with selective reflecting mirrors. Two lamps have been developed with the following reflectance properties:

HER-lamps 12 V 75 W	Maximum reflectance	Reflectance of 50 %
Ultraviolet mirror	(375 ± 15) nm	(290 ± 20) nm (420 ± 20) nm
Blue mirror	(420 ± 20) nm	(370 ± 20) nm (530 ± 20) nm

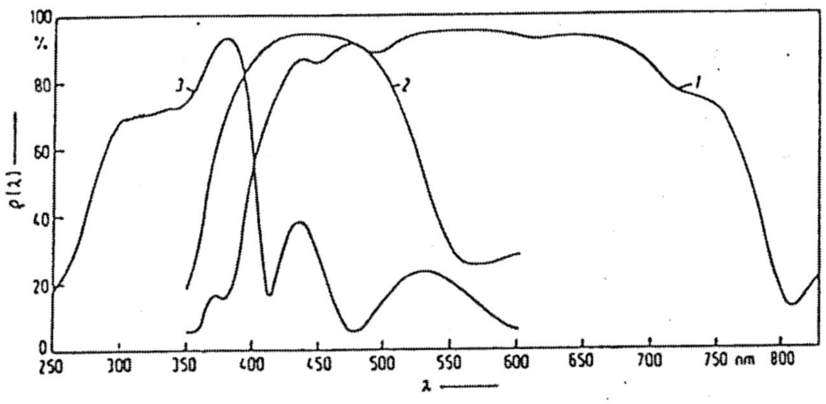

Fig. 1 Spectral reflectance curves

1 cold mirror **2** blue mirror **3** ultraviolet mirror

Typical reflectance curves of those ultraviolet and blue mirrors are shown in Figure 1 in comparison with the reflectance of a cold mirror.

3. Measurement of the radiation

The luminous flux of HER-lamps with blue mirrors is measured with reference to DIN 15 749, weighted according to the spectral luminous efficiency function.

A special detecting system with a peak sensitivity at 350 nm is used to measure the irradiance of HER-lamps with ultraviolet mirrors in the range between 250 nm and 400 nm, typical irradiance values are about 200 W/m^2.

The lamp types HER 12 V 75 W with blue mirror (64613) and HER 12 V 75 W with ultraviolet mirror have been designed for an average lifetime of 25 hours. The decrease of the luminous flux or ultraviolet irradiance, respectively, during lifetime is comparable with cold mirror HER-lamps.

DIAGNOSTICS

46 - 62

DETERMINATION OF TWO-DIMENSIONAL TEMPERATURE AND ADDITIVE DENSITY DISTRIBUTIONS IN A METAL-HALIDE LAMP ARC

G.L. Rogoff, A.E. Feuersanger, and J.P. Drummey

GTE Laboratories Inc., Waltham, Massachusetts 02254, U.S.A.

H.L. Rothwell, Jr.

GTE Lighting Products, Salem, Massachusetts 01970, U.S.A.

Segregation of additive species can lead to significant axial and radial nonuniformities in the electrical and radiative characteristics of a vertical high-pressure arc. An experimental method has been developed to determine both the radial and axial distributions of temperature and additive density in a rotationally symmetric LTE arc. This method can be used if one knows the total vaporized quantity of the major component of the gas mixture in the arc tube. The pressure is determined along with the distributions.

The procedure consists essentially of the following steps: transverse scans of emission from the major (buffer) species and additive species of interest are made at several axial locations;[1] the measured absolute intensities are converted to radial distributions of emission coefficients; the buffer emission coefficients are used to determine the pressure, which is taken to be uniform, and the radial temperature distributions at the selected axial locations (an iterative process is used here); and the resulting temperature profiles are utilized along with the additive emission coefficients to determine the additive density profiles at those same axial locations. If the emission scans are made in optically thin spectral regions, an Abel inversion can be used to obtain the emission coefficients from the intensity scans. The transverse intensity scans can be recorded by various means, e.g., by electronic imaging devices coupled to computers to allow rapid implementation of the procedure and for time-resolved measurements. In the demonstration of the procedure described here, the intensities were obtained from densitometer scans of filtered arc photographs.

The method has been demonstrated by applying it to a 175 W Na-Sc-Hg arc, with mercury the major component and sodium the additive of interest. The arc, which is similar to that in a standard 175 W Sylvania Metalarc® lamp, is operated at 60 Hz. In steady-state operation the predetermined mercury dose is totally vaporized, fixing the total number of mercury atoms in the vapor, while a condensate of the additives remains. The distributions are determined from side-on time-integrated photographic arc images of mercury emission through a narrow band filter at 577 nm, and of sodium emission through a filter at 593 nm in the red wing of the D line, both spectral regions being reasonably optically thin. To relate the D-line emission to the sodium density we use an approach[2] developed to describe alkali-rare gas excimer emission. Transverse calibrated densitometer scans are made at four axial locations on each film image, with the output digitized for computer data processing. For the mercury scans, a modified parabola is used to extend the temperature distributions to the wall.

The resulting profiles clearly show the effects of segregation. The upper portion of the arc and the axial regions are significantly depleted of sodium.

1. A similar procedure for an axially uniform temperature distribution is described in R.J. Zollweg, J.J. Lowke, and R.W. Liebermann, J. Appl. Phys. 46, 3828 (1975).
2. R.E.M. Hedges, D.L. Drummond, and A. Gallagher, Phys. Rev. A 6, 1519 (1972).

SPECTRAL LINE BROADENING OF THE MERCURY 365 nm AND 405 nm LINES IN PULSED OPERATION OF AN HBO 350 W/CC.

G. Lüthen, W. Steudtner
OSRAM GmbH, D-8000 München 90

Introduction:

For the exposure of photo resists pulsed operation of short arc mercury lamps using, e.g., the Hg-lines 365 nm or 405 nm can be employed. High spatial resolution requires narrow filtering (e.g. ± 3 nm for the 365 nm line and ± 5 nm for the 405 nm line) in this application. Consequently, the relatively strong broadening of the 365 nm line would lead to significant losses in radiant intensity, whereas with the weaker broadening of the 405 nm line, no such losses should occur. It was the aim of this work to determine the behaviour of the unfiltered and filtered profiles of these lines using an HBO 350 W/CC under pulsed operation.

Experimental Set-up:

The lamp was operated in a 2 Hz cycle with a pulsed power supply of MIMIR-Corporation (St. Clara, Cal.,USA), controlled by a photodiode: if this diode was illuminated, the output power rose to 500 W; after 200 ms the incident light was shut off by a chopper and the output power was decreased to the idle level of the supply (300 W) for 300 ms. Furthermore there was a 1.1 kW spike at the very beginning of the pulse followed by an 800 W plateau, 40 ms long. With two fused silica lenses an enlarged image of the lamp was generated in the plane of the entrance slit of a 1m JARRELL-ASH-spectrometer of Czerny-Turner-mounting. The line profiles were measured time resolved with an optical multichannel analyzer (OMA-3 system, EG&G-PARC, Munich); the intensified diode array camera of the OMA was mounted at the exit port of the spectrometer.

The OMA was triggered with the 1.1 kW spike of the lamp pulse. The delay unit in the gate pulse generator allowed measurements at different times of the lamp pulse using a camera exposure time of about 100 µs. Thus the temporal development of the line profiles across the lamp pulse was determined.

<u>Results:</u>

* The variations in radiant intensity and line profiles start within 1 ms after power variation.

* The full widths at half maximum (fwhm) of the Hg-lines are:

Lamp power:	300 W	500 W	800 W
365 nm:	4.84 nm	7.10 nm	9.51 nm
405 nm:	2.00 nm	2.35 nm	2.61 nm

* Comparing the integrals of the two Hg-lines for the full widths of the lines at half maximum against the widths of the filter transmission curves shows for the 365 nm line practically no loss of radiant intensity at 300 W, a loss of ca. 25% at 500 W and about 50% at 800 W due to cutting off the wings of the strongly broadened line. For the 405 nm line practically no loss in radiant intensity could be found.

TRANSIENT BEHAVIOUR OF THE PLASMA IN THE ELECTRODE REGION IN HPS-LAMPS STUDIED BY HIGH-SPEED CAMERA AND EMISSION SPECTROSCOPY

J.Szántó, S.Holló and I.Újj

TUNGSRAM Ltd.Co. 1340 Budapest Váci út 77. Hungary

The operation of a HPS-lamp is affected by some of the processes, which are going on as long as the steady state plasma develops. Our aim was to study the most critical phenomenon, the removal of the activator from the electrode during the ignition.

The ignition of 400 W HPS-lamps operated by 50Hz AC was examined. Transparent sapphire discharge-tube and Ba_2SrWO_6 activator was used. The time-dependent variation of the lamp-current, -voltage and the intensity of the 455.4nm Ba line were registered and the electrode region of the discharge was recorded by high-speed camera simultaneously. (10000 frame/sec film speed was used.)

Our findings can be summarized as follows:

1.The cathode-half-period begins with the luminous Xe-light of the glow discharge which covers the cathode. Its duration correlates with the width of the reignition peak in the lamp-voltage. When the increasing current cannot be maintained any more by the glow, it abruptly collapses and a quickly-moving-spot (QMS) state occurs. Later, when the electrode is warmer, a stabilised thermoemission arc follows the glow, which at higher currents, is transformed to be a QMS-state again. In the rest of the half-period the plasma remains in QMS-state. Sometimes the reoccurance of the glow can be observed at the very end of the cathode-half-period.

2.During the glow and abnormal glow phase, when the cathode fall and therefore the energy of the bombarding Xe ions is much higher than in the QMS-state, removal of the activator cannot be detected. Ejection of the activator appears solely during the QMS-state, so it cannot be explained by the sputtering effect of the impinging Xe ions as usually interpreted [1]. The qualitative explanation for this is the following: the diameter of the spot is very small so the electron current density has to be very large to yield the appropriate current. As a consequence of the low conductivity of the cathode activator its local temperature suddenly rises up due to the Joule-heating and the rapid thermal expansion induces the explosive ejection of the activator from the area where the moving spot just has been staying. The high electron current density can give rise to electrolytic processes in the oxide-based activator as well. Electrostatic and polarisation effects should also be concerned with. As a result not only atoms and ions but also small pieces of the activator material are removed from the cathode.

At the very beginning of the next anode-half-period an intensive streaming of Xe gas delivers the ejected activator material to the tube wall. The slower diffusion has much less role in this process.

Acknowledgement: We are grateful to Mr.F.Nagel and B.Nyiri for many valuable discussions.

Ref.:1. Hirayama et.al. Journal of IES January 1981. p.90

Nr. 49

REAL-TIME SPECTROSCOPIC MEASUREMENT OF TEMPERATURE PROFILES IN LTE-LAMPS

M.Brandenbusch and W.Heering
Lichttechnisches Institut, University of Karlsruhe, Kaiserstr.12
Federal Republic of Germany

Dedicated to Prof. Dr. P. Schulz on the occasion of his 75th birthday

There are a lot of methods available to determine temperature profiles in lamps. But most of them do not qualify for on-line measuring local temperature profiles as a function of time. We propose a way to do this as well on optically thin as thick plasmas without preliminary knowledge of pressure and atomic quantities except that of excitation energies.

The two procedures that we are realizing with fast spectroscopic multichannel analysis combine or modify known methods of temperature determination in cylindrically symmetric plasmas. If there are some optically thin spectral lines emitted from the plasma that originate from different energy levels, broad-band spectra are measured simultaneously at several line-of-sight positions and up to one hundred of these two-dimensional spectra are recorded closely one after the other. Then, after fast Abel-inversion of each set of line intensities, the temperatures of respectively two concentric plasma rings are calculated from the functional relation between the ratio of the line emission coefficients of these two zones and the associated excitation energy.

If the lamp plasma is not optically thin, one measures the two-dimensional spectrum of the lamp and very shortly after, that of the lamp with a back mirror that is then swept in the direction of sight. From the two measured spectral radiances along the line-of-sight the average spectral absorption coefficients of the different temperature rings can be iteratively computed. Then, applying Kirchhoff's law to each temperature zone the Planck function of each ring and thus its temperature will be determined by means of another iteration. For this procedure, the absolute values of spectral radiance have to be measured. We do not perform this as usually by comparing with the spectral radiances of a radiation standard whenever the temperature profile of a lamp is to be determined, but we sense the absolute irradiance by a calibrated GaAsP photodiode with an arched teflon diffuser in front of it and very near the entrance slit of the polychromator just at the time of spectral detection.

For the spectral measurements, we use a holographic concave grating of flat field type with blaze wavelength at 250 nm and two matrix arrays of 256x256 silicon photodiodes with response maxima at 750 nm as image sensors. Because of the relatively small radiances of the lamp boundary we intensify by a two-stage proxifier in front of the arrays. Its screen radiances are fibreoptically coupled to the arrays. The part of its photocathode that receives the visible part of the spectrum is covered by a filter foil to suppress the second order of diffraction. Thus, the broad spectral range from 200 nm to 800 nm that is necessary for good Boltzmann plots can be detected by only one scan. As only a few lines are of interest each time, we do not read out the total frame but only small fields of it. To do this we have provided a control system that obtains the gating signals for the solid state cameras from a programmable random-access-memory. The control program will be made according to the spectral parts to be detected and loaded into the RAM from a PC via a PIO-device before spectral sensing. This will largely speed up the spectral acquisition. For still higher speed spectra processing we want to accumulate a programmed number of digitized video signals from neighbouring columns on-line. This will raise the signal-to-noise ratio, but limit the local resolution. The Abel-inversion will be performed nearly in real-time by fast matrix multiplication with each set of acquired spectral line intensities from different positions. A word-slice processor system is just being developed for fast accumulation and matrix multiplication.

DETERMINATION OF THE EXCITED STATE DENSITY FOR AN OPTICALLY THICK LINE

R.W. Richardson* and S.M. Berman

Lighting Systems Research Group, Applied Science Division
Lawrence Berkeley Laboratory, University of California
Berkeley, California, U.S.A. 94720

The radial distribution of excited-state atoms is an important characteristic of cylindrical discharges such as the Hg-Ar discharge used in fluorescent lighting. These distributions can be determined from absorption measurements using a transition from the state of interest to a higher excited state if the discharge is optically thin for the line chosen. They can also be determined from the emission of an optically thin line. In these cases, the dependence of the absorption or emission on the distance from the axis is measured (see figure). The radial dependence of the density of excited-state atoms is then determined by inverting this data using a solution of Abel's integral equation[1]. This inversion is only rigorous in the limit of vanishing optical thickness, a condition often difficult to produce in realistic conditions. We report the solution of the corresponding integral equation for arbitrary optical thickness. This can be used as a check on the use of the Abel solution, as an approximation for finite optical thickness, or it can be used as a direct method of determining the excited-state density of an optically thick resonance line.

The emission intensity of a non-overlapping line at a fixed optical thickness Λ, i.e., fixed frequency, and at a radial offset ρR, where R is the radius of the discharge (see figure), is given

$$I(\Lambda,\rho^2) = \Lambda \int_{-\xi_o}^{\xi_o} d\xi \, e^{-\Lambda(\xi_o - \xi)} f(\rho^2 + \xi^2) \tag{1}$$

where $\xi_o = (1 - \rho^2)^{1/2}$ and f is the radial distribution of excited-state atoms giving rise to the emission, and we have assumed that there is no scattered contribution to the intensity.

This work was supported by the Assistant Secretary for Conservation and Renewable Energy, Office of Buildings Community Systems, Building Equipment Division of the U.S. Department of Energy under Contract No. DE-AC03-76SF00098.

*Permanent Address: New York University, New York, NY, U.S.A. 10003

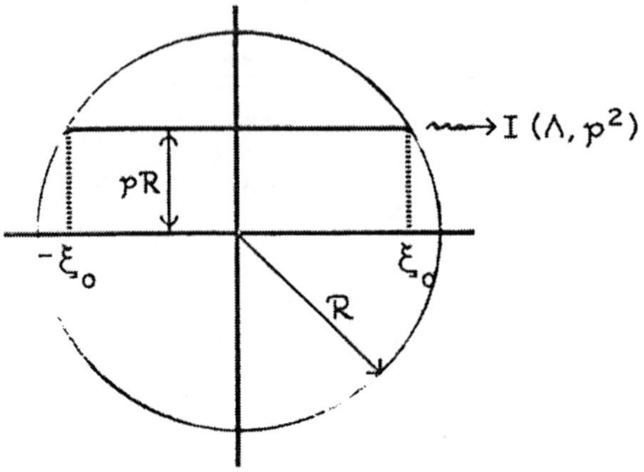

Cross Section of Discharge

If we define $q = \rho^2$, $x = \rho^2 + \xi^2$, and $F(\Lambda,q) = e^{\Lambda \xi_o} I(\Lambda,q)/\Lambda$, then (1) becomes:

$$F(\Lambda,q) = \int_q^1 dx \frac{\cosh[\Lambda (x-q)^{\frac{1}{2}}]}{(x-q)^{\frac{1}{2}}} f(x). \qquad (2)$$

Using methods similar to those of Abel[1], we have been able to invert (2) with the result

$$f(x) = -\frac{1}{\pi} \int_x^1 dq \frac{\cos[\Lambda (q-x)^{\frac{1}{2}}]}{(q-x)^{\frac{1}{2}}} \frac{\partial F(\Lambda,q)}{\partial q}. \qquad \Lambda \geq 0 \qquad (3)$$

This can also be verified by substituting (2) and (3) and showing the result to be an identity. This solution is valid for arbitrary Λ and reduces to the standard solution of the Abel equation in the limit $\Lambda \to o$. A determination of f from a measurement of total intensity of the line as a function of q would also be useful. This involves putting a line shape into (1), i.e., Λ as a function of frequency e.g., a Dopper profile, and integrating it over frequency. We have not been able to generate a closed-form solution such as (3) for the resulting equation. However, techniques for determining f from such data will be presented and used to analyze observational data.

Reference:

(1) E.T. Whittaker and G.N. Watson, Modern Analysis, Cambridge University Press, 1952, p.229.

MEASURING HIGH TEMPERATURE ABSORPTION COEFFICIENTS

D O Wharmby, Research and Engineering Laboratories, THORN EMI Lighting Limited, Melton Road, Leicester LE4 7PD, England

N R Ede, Department of Physics, Loughborough University of Technology, Loughborough, England

J Meads, Department of Chemistry, University of Sheffield, Sheffield, England

Absorption coefficients $\kappa(\nu,T)$ of the high temperature vapours in LTE arcs (at temperatures T and frequency ν) are the input data for radiation transport calculations. They are therefore essential for adequate modelling of the spectral properties of HID lamps. As yet, satisfactory values of $\kappa(\nu,T)$ are only available in a few cases. For example, resonance broadening has been used successfully to calculate $\kappa(\nu,T)$ for the sodium D lines in HP sodium arcs. Even this fails at sodium pressures $\geq$100 torr because of inadequate line broadening theory.

We report here a method for extracting the value of $\kappa(\nu,T)$ from measurements of the spectral radiance of an LTE arc. The absorption coefficient for colliding atoms A and B can be expressed[1,2] in a very simple way as

$$\kappa(\nu,T) = n_A \; n_B \, a(\nu) \; \exp(-b(\nu)/T) \tag{1}$$

where n_A and n_B are atom densities and $a(\nu)$ and $b(\nu)$ are constants which may be determined from the properties of the molecule AB in the few cases that they are known[2].

Using the radiation transport equation the *calculated* radiance of an LTE plasma, diameter 2X, having a temperature profile $T_i \equiv T_i(x)$ is then given by[3]

$$I_\nu^c(i)=\int_{-X}^{X} n_A(T_i)n_B(T_i)a(\nu)\exp(-b(\nu)/T_i)B_\nu(T_i)\exp\left[-\int_{x}^{X} n_A(T_i)n_B(T_i)\,a(\nu)\,\exp(-b(\nu)/T_i)\,dx'\right]dx \tag{2}$$

where $B_\nu(T_i)$ is the Planck radiance. We now *measure* the spectral radiance $I_\nu^m(i)$ for i=1 to N temperature profiles $T_i(x)$. The constants $a(\nu)$ and $b(\nu)$ can then be obtained at each frequency by minimizing the quantity

$$\sum_{i=1}^{N}\left[\,1-\frac{I_\nu^m(i)}{I_\nu^c(i)}\,\right]^2 \tag{3}$$

using a suitable non-linear optimization method.

Numerical simulations of typical measurements show that the constants $a(\nu)$ and $b(\nu)$ can be recovered from expression (3) under a range of conditions which might apply in an experiment. Best results are achieved by making measurements under conditions in which the maximum temperature is altered over the widest possible range. Provided that the temperature profile can be accurately measured, high temperature (>2500K) absorption coefficients are retrieved with an accuracy of better than 1%. At lower temperatures, there is increasing error, principally because these regions contribute so little to the radiance. Simulations show that spectral radiance measurements must be made with very high precision. The effect of measurement errors in the temperature profile used to evaluate equation 2, have also been investigated.

In making measurements, the radiance values needed can be obtained by making use of the varying temperature profiles which occur during the 50Hz AC operation of the arc. A wider range of temperatures may be obtained by switching off the arc periodically. High precision (S/N ratio ~1000 at the dimmest part of the cycle) has been achieved by sampling the radiance 100 times per cycle for periods of 100μs and for typically 100 cycles. The computer controlled equipment for doing this large number of measurements will be described. Arc temperatures are measured by the method described by Cifuentes[4].

Measurements of some absorption coefficients in the extreme wings of the sodium D lines will be presented. These will then be used in the radiation transport equation to predict emission spectra. Possible applications in obtaining data for metal halide arcs will be considered.

1. Hedges R E M, Drummond, D L and Gallagher, A, Phys. Rev. A, 6(4), 1519-1544, 1972.
2. Wharmby D O, "Molecular spectra in LTE plasmas", presented at the NATO Advanced Study Institute, Pitlochry, Scotland July 1984, to be published.
3. Richter J, "Radiation from hot gases", in "Plasma Diagnostics", ed. W. Lochte-Holtgreven, (Amsterdam:North Holland), 1968.
4. Cifuentes L, "Temperature measurement using self reversed NaD lines", this conference.

APPLICATION OF DIFFERENTIAL QUADRATURE TO THE SOLUTION OF THE ENERGY BALANCE EQUATION IN HIGH PRESSURE LAMPS

P. A. Vicharelli and L. C. Pitchford
GTE Laboratories, Inc.
40 Sylvan, Rd., Waltham, MA 02254 USA

The spatial temperature profile of a high pressure arc lamp may be obtained by solving the energy balance equation. For the case of a cylindrically symmetric, axially uniform arc, this equation takes the form

$$\rho C_p\left(\frac{\partial T}{\partial t}+v_r\frac{\partial T}{\partial r}\right)=\sigma E^2-U_R+\frac{1}{r}\frac{\partial}{\partial r}\left(r\kappa\frac{\partial T}{\partial r}\right), \qquad (1)$$

where $T(t,r)$ is the temperature, r is the radial coordinate, v_r is the radial velocity, C_p is the specific heat at constant pressure, κ is the thermal conductivity, σ is the electrical conductivity, E is the electric field strength, ρ is the density, and U_R is the local net emission coefficient. The radial velocity may be calculated from the mass continuity equation

$$\frac{\partial \rho}{\partial t}+\frac{1}{r}\frac{\partial}{\partial r}\left(r\rho v_r\right)=0, \qquad (2)$$

and the field strength is calculated according to Ohm's law,

$$E = I/2\pi\int_0^R \sigma\, r\, dr, \qquad (3)$$

where R is the tube radius. Values for the material properties ρ, C_p, σ, and κ must be known as functions of temperature. The usual approach is to tabulate these quantities for a few temperatures and interpolate as values are needed. The net emission coefficients U_R are obtained by solving the radiation transport equation.

The purpose of this paper is to present a simple method for solving the energy balance equation shown above. The method is based on differential quadratures,[1,2] where a derivative of order m can be expressed by

$$\frac{\partial^m f(r_i)}{\partial r^m}\simeq\sum_{j=1}^{N}W_{ij}\,f(r_j), \qquad (4)$$

where the r_i, $i=1,2,\ldots,N$, are the grid points, the $f(r_i)$ are the function values at these points, and the W_{ij} are the weights associated with these function values. Equation(4) is exact for polynomials of degree less than or equal to (N-1). When the spatial derivatives in Eqs.(1) and (2) are replaced by sums as in Eq.(4), the result is a set of ordinary differential equations. In the case of a steady-state problem a set of algebraic equations is obtained. In both instances established numerical methods are available for solving the resulting set of equations. The weight coefficients are calculated by analytically inverting a Vandermonde matrix.[3] This is done only once before the calculation begins. For the m=0 case the weights may also be used to interpolate the material properties and to evaluate the electric field integral. In other words, the same method may be used to uniformly handle the various phases of the problem.

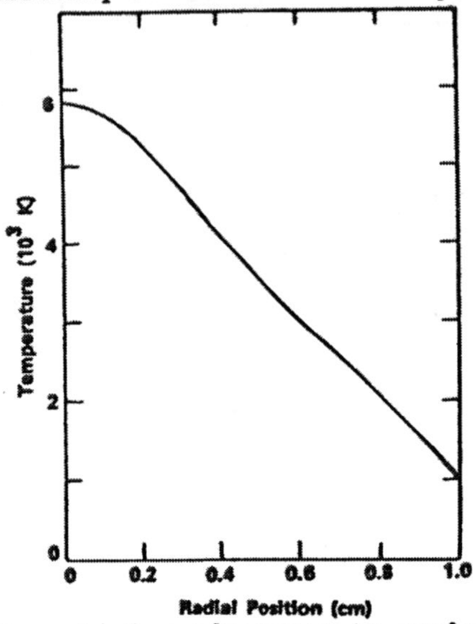

Fig. 1 Radial temperature profile for a mercury arc with I=1 A, p=1 atm, and R=1 cm.

Using differential quadratures to evaluate derivatives, the computational effort is minimal relative to other methods[4-7] because only a few grid points (3 to 5) are all that is usually needed. An added advantage of the present method over orthogonal collocation[6] is that the location of the grid points is arbitrary and may be chosen to cluster around interesting regions of the temperature profile. An adaptive grid is straightforward to implement, as is the generalization to multidimensional problems. As an illustration of the method, typical results of a calculation for a high pressure Hg discharge are presented in Fig. 1. The material functions are those used by Lowke[5] and, for simplicity, we take U_R to be another material property parameterized as a function of temperature.

REFERENCES

1. R. E. Bellman and J. Casti, J. Math. Anal. Appl. 34, 235 (1971).
2. F. Civan and C. M. Sliepcevich, J. Math. Anal. Appl. 93, 206 (1983).
3. R. W. Hamming, "Numerical Methods for Scientists and Engineers," 2nd ed., McGraw-Hill, New York, 1973.
4. J. J. Lowke, R. E. Voshall, and H. C. Ludwig, J. Appl. Phys. 44,3513(1973).
5. J. J. Lowke, J. Appl. Phys. 50, 147 (1979).
6. H. E. Lee, J. Phys. D 16, 2191 (1983).
7. H. P. Stormberg and R. Schafer, J. Appl. Phys. 54, 4338 (1983).

INTENSITY OF SELF-REVERSED LINES

P. A. Vicharelli and L. C. Pitchford
GTE Laboratories, Inc.
40 Sylvan, Rd., Waltham, MA 02254 USA

Information about the spatial characteristics of plasmas may be obtained through spectroscopic measurements of emitted line intensities. In high pressure arc lamps a considerable amount of re-absorption is usually present and the spectra become self-reversed. A method for analyzing this self-reversal effect was proposed by Bartels[1,2] some time ago. The intensity is expressed in terms of a lineshape function

$$Y_\lambda(\tau_T, p) = e^{-\frac{\tau_T}{2}}\left[\frac{\tau_T}{2}(1-p) + p\sinh\left(\frac{\tau_T}{2}\right) + \frac{1}{p^{1/2}}\sinh\left(\frac{\tau_T}{2}p^{1/2}\right)\right], \qquad (1)$$

where τ_T is the total optical thickness along the line of sight and p is a parameter proportional to the second moment of the source function. Unfortunately, this function fails when τ_T becomes large.

The purpose of this paper is to present a simple one-parameter model for the intensity of self-reversed lines emitted from an inhomogeneous plasma of arbitrarily large τ_T. An expression for the observed side-on intensity, I_λ, along a particular line of sight is obtained by analytically integrating the radiative transfer equation in one dimension. The result may be expressed in the form $I_\lambda = CL_\lambda(\tau_T)$ where C is a term independent of wavelength, and $L_\lambda(\tau_T)$ is a lineshape function. The lineshape function is given by

$$L_\lambda(\tau_T) = \tau_T e^{-\frac{\tau_T}{2}} \int_{-1/2}^{1/2} \sigma(\xi)\, e^{\tau_T \xi} d\xi, \qquad (2)$$

where $\sigma(\xi)$ is a normalized source function, and ξ is the optical thickness in units of τ_T.

Next, the integral above is evaluated using a simple model source function constructed from the weighted average of 1) a uniform term of width W and height H, and 2) a monotonically decreasing inhomogeneous term that vanishes at the boundaries. The uniform term is characterized by a parameter that is fixed by matching with an asymptotic expansion[3] of the emission profile in the limit of large optical thickness. The weight of the inhomogeneous term is found by requiring that the first moment of the reduced source function be normalized to unity. Our resulting lineshape is given by

$$L_\lambda(\tau_T, W) = e^{-\frac{\tau_T}{2}}\left[2W\sinh\left(\frac{\tau_T}{2}W\right) + (1-W^2)\tau_T\right]. \qquad (3)$$

We then consider the extension of this result to arbitrary sources. As in the method of Bartels, we characterize the source with the parameter p. We obtain

$$L_\lambda(\tau_T, p) = e^{-\frac{\tau_T}{2}}\left[2p^{1/4}\sinh\left(\frac{\tau_T p^{1/4}}{2}\right) + \left(1 - p^{1/2}\right)\tau_T\right]. \tag{4}$$

This lineshape is plotted in Fig. 1 for various values of p. Also shown for comparison is a set of curves calculated using Eq.(1). As we can see in this figure, the present model agrees rather well with that of Bartels for low and moderate values of τ_T. Discrepancies are observed in the region of larger τ_T because Eq.(1) begins to fail there, while Eq.(4) remains valid in this limit because the proper asymptotic behavior is built-in.

As a test, we have calculated a self-reversed lineshape for a case that could be handled analytically, namely, a parabolic source function. We also evaluated the corresponding p parameter and calculated the lineshape predicted by Eqs.(1) and (4). The accuracy of the predictions is illustrated in Fig. 2.

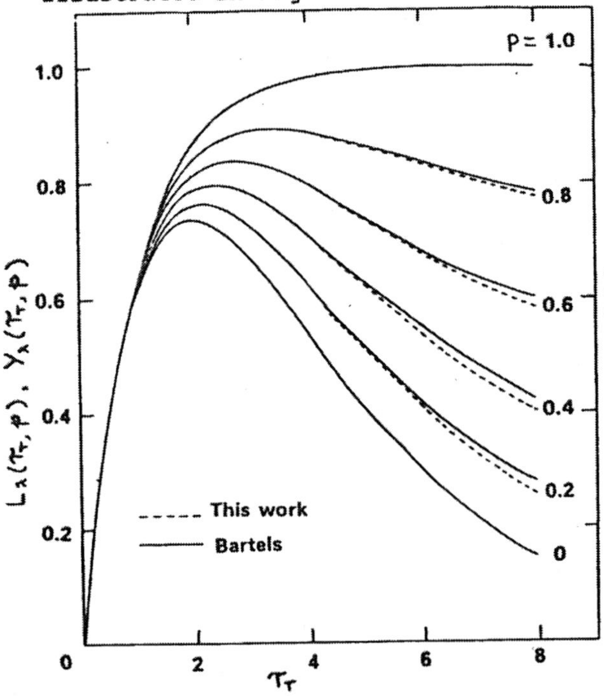

Fig. 1 One-parameter self-reversed lineshapes.

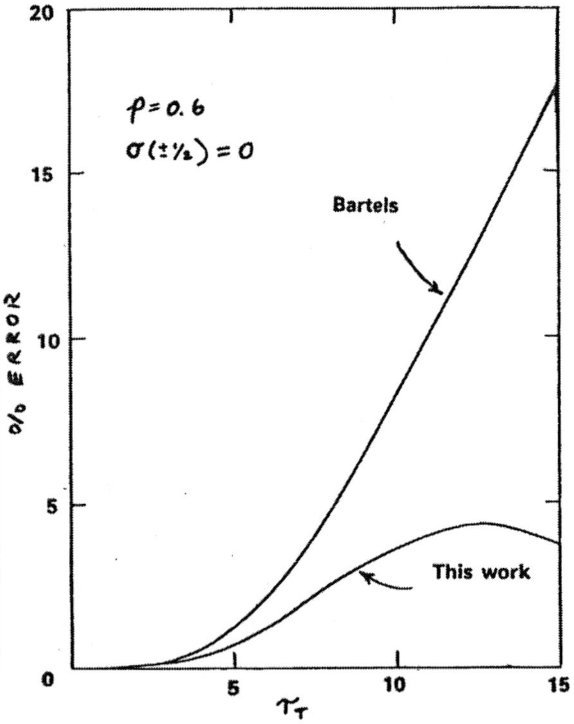

Fig. 2 Accuracy of the one-parameter lineshapes applied to a parabolic source.

REFERENCES

1. H. Bartels, Z. Phys. 127, 243 (1950).
2. H. Bartels, Z. Phys. 128, 546 (1950).
3. P. A. Vicharelli, Paper DB-5, 38th GEC, Monterey, CA, 1985.

ARC TEMPERATURE MEASUREMENT USING SELF-REVERSED Na D LINES

L Cifuentes, THORN EMI Lighting Limited, Research and Engineering Division
Melton Road, Leicester LE4 7PD, U.K.

Arc temperatures in lamps are often measured by Bartels' method using the 818 nm sodium lines[1,2]. In many cases this is not possible, e.g. when measuring temperature near current reversal in a high pressure sodium lamp, as then the 818 nm lines are no longer self-reversed. Similar problems are likely to arise in metal halide arcs including sodium halides. By using a Bartels' like method, the sodium D lines provide a viable alternative, provided that the right numerical method and physical assumptions are applied.

Radiation transport calculations on a large variety of conditions have shown that the peak intensity of a self-reversed line depends strongly on centre temperature, weakly on temperature profile and even less on pressure.

Spectral intensity calculations were carried out using an expression valid for LTE steady state arcs[3]

$$I_\nu = \int_{-R}^{R} K(\nu,T)\, B_\nu(T) \exp\left[-\int_{x}^{R} K(\nu,T)dx'\right] dx \qquad \text{(A)}$$

where the units of I_ν are W m^{-2} sr^{-1} nm^{-1} and

$K(\nu,T)$ = absorption coefficient $B_\nu(T)$ = blackbody function

R = arc tube radius   x = distance along the line of sight

The absorption coefficient $K(\nu,T)$ was calculated assuming resonance broadening, the expressions for the semi half widths being taken from reference 4. The sodium number density was calculated as a function of the arc tube radius by a Gibbs free energy minimization method allowing for de-mixing, and the semi half widths were evaluated as a function of the sodium number density. The arc temperature profile was taken to be

$$T = T_C - (T_C - T_W)\left(\frac{x}{R}\right)^{\alpha}$$

where T_C = centre temperature, α = shape factor and T_W = wall temperature. de Groot[5] finds the shape factor to be between 2 and 2.5 experimentally, but we have used values between 1.5 and 3.5 for these calculations.

Equation (A) was evaluated for a series of centre temperatures and with a wide range of values for T_w, α, R and n. The quantities $\Delta\lambda_r$ and $\Delta\lambda_b$ (the separation between the red and blue wing self-reversal peaks and the corresponding D line average wavelengths) and $I_{max,r}$ $I_{max,b}$ (the intensities of the self-reversal maxima) were obtained from the calculation. This allowed a relationship for the dependence of the centre temperature on lamp parameters of the form

$$T_c = f(I_{max}, T_w, n_{Na}, \alpha, R)$$

to be established for future temperature measurements. The effect of adding mercury on the red wing width was then investigated. The possibility of extending this procedure to arcs containing sodium halides has also been studied.

Measurements of the intensity of Na D lines and the 818 nm line in transparent sapphire tubes near the maximum temperature in a 50 Hz ac arc show agreement between the two methods.

Our results allow a comparison to be drawn between Bartels' method on the 818 nm lines with the D lines method. The main conclusions are that the intensity of the self reversed sodium D lines can be used to find the maximum temperature along a line of sight through a high pressure LTE sodium plasma. The approximate form of the temperature profile must be known. If there is little error in this factor, then the accuracy of the D line method is likely to be better than the Bartels' 818 nm line method, currently thought to be the best. Both methods require self reversal, so the D line method has much wider applicability, and can be applied to metal halide arcs.

(1) J J de Groot and A G Jack, J Quant.Spectrosc.Radiat. Transfer, 13(1973),615-626.

(2) H Zwicker in 'Plasma Diagnostics' (W.Lochte-Holtgreven, Ed.), North-Holland, Amsterdam, 1968, pp.214-249.

(3) J Richter in 'Plasma Diagnostics' (W.Lochte–Holtgreven, Ed.), North-Holland, Amsterdam, 1968, pp.1-65.

(4) B F Jones and D A J Mottram, J.Phys.D: Appl.Phys., 14(1981), 1183-94.

(5) J J de Groot, Ph.D. Thesis, Technical University of Eindhoven, 1974.

AN ELEMENTARY MODELLING OF THE RADIATIVE FLUX EMITTED FROM A SELF-REVERSED LINE USING BARTELS'S FORMALISM

J.J. DAMELINCOURT, M. AUBES, T. ABDENNADHER and N. SALERNO
Laboratoire des décharges dans les gaz, UA 277 du CNRS, Toulouse, CPAT

E. ROAUX
Service recherche de la Compagnie des Lampes et du Matériel d'Eclairage, LIMEIL BREVANNES, France

Semi empirical modelling of high pressure discharge, assuming known parabolic temperature profiles, may be of interest in lamp design, particularly in evaluating the contribution of each line to the lumen output. This work describes a simple modelling of the radiative flux emitted from a self absorbed line ; BARTELS' formalism is used in performing the model. For two lines (Hg 456.1 and Tl 535 nm) a comparison is made of the actual computed results with those obtained from the complete resolution of the equation for radiative transfer. In BARTELS' formalism /1/ the total luminance of a line was written in L.T.E. conditions under the form

$$L_T = \int_{-\infty}^{+\infty} L(\lambda)\, d\lambda = MW(\lambda_o, T_o) \int_{-\infty}^{+\infty} Y(\tau, p)\, d\lambda \qquad (1)$$

Where $W(\lambda_o, T_o)$ denotes the maximum value of the WIEN'S function along the line of sight and the function $Y(\tau, p)$ represents the influence of the optical thickness. Using τ in place of λ in the integral one can write /2/;/3/

$$L_T = W(\lambda_o, T_o).M.A^{1/2}\, \delta^{1/2}\, S(\tau_{max}) \qquad (2)$$

$$A = \frac{\lambda_o^4}{8\pi C} A_{ea} \frac{g_e}{g_o} \frac{p_a}{k} \frac{1}{\pi} \int_{-x_o}^{+x_o} \frac{1}{T(x)} \exp \frac{-E_a}{kT(x)}\, dx \qquad (3)$$

p_o is the partial pressure of the emiting gas, $S(\tau_{max})$ depends slightly on τ_{max} for τ_{max} greater than 5. According to equation 2 it is possible to calculate the intensity emitted by a radial plane of unit depth in a direction θ

$$(\xi I_T)_\theta = (\cos\theta)^{1/2} \times K(\xi, M, \beta)\, L_T(g=0, \theta=0) \qquad (4)$$

where $K(\xi, M, \beta)$ is a slowly varying function of M, β and ξ if $T' = \xi(1-(1-1/C)\rho^\beta)$ and $\xi = kT_o/E_i$.

The flux, in the limits of the assumptions made, may be easily deduced from (4) :

$$F = 4\pi\, K(\xi, M, \beta) L_T(g=0, \theta=0) \int_0^{\pi/2} \cos^{3/2}\theta = d\theta = 10.98\, K(\xi, M, \beta) L_T(0,0)$$

Furthemore the equation for radiative transfer has been solved in terms of luminance and flux for two lines ; the Hg 546.1 nm line in pure Hg discharges and the Tl 535 nm line in HgTl discharges. The lamps studied were 10 cm long and 2 cm in diameter.

TABLE 1 - Flux values for 10 cm of discharge ; from equation for radiative(F_R) ; from the actual formalism and the theoretical values of $\delta(F_B)$.The values marked (*) are calculated with δ_{eff} values obtained from measurement of 2S the distance between line maxima /3/.

Line		Hg 546,1 nm				Tl 535 nm	
Lamp power (W)		1030	1030	640	640	1030	1030
Hg pressure (torr)		1824	3500	1670	3120	1140	3340
$Tl^+ + Tl^n$ pressure (torr)		0	0	0	0	7.7	10.2
t = 5 ms	F_B (W)	52	58	28	34	53/70*	67/74*
	F_R (W)	47	57	24	33	69	83
t = 0 ms	F_B (W)	12	18	6.2	11	25/30*	41/43*
	F_R (W)	15	22	7.5	13	31	48

This work, sustained by the "Club EDF lumière et vision", was performed as a part of a joint research program between "Centre de Physique Atomique/UA 277 du CNRS" and "service recherche de la Compagnie des Lampes et du Matériel d'Eclairage".

REFERENCE

/1/ LAPWORTH K.C., Nat. Phys. Lab. Report Q55 1980.
/2/ FUNK W. and KLOSS H.G., Beitr. Plasmaphysik 13 101 1973.
/3/ ABDENNADHER T., Thèse de spécialité n°3237 Toulouse 1985.

BROADENING AND SHIFT OF THE 655 nm THALLIUM LINE EMITTED BY A HIGH PRESSURE Hg-TlI DISCHARGE

J.J. DAMELINCOURT, M. AUBES, T. ABDENNADHER
Laboratoire de décharges dans les gaz, UA 277 du CNRS, CPAT,
118 route de Narbonne, 31062 TOULOUSE CEDEX, France,
D. KARABOURNIOTIS, S. COURIS,
Department of Physics, University of CRETE, IRAKLION, Crète, Grèce

Line profile analysis is a powerful tool for discharge diagnostic. In this paper we report on broadening of the $8^2P_{3/2}-7^2S_{1/2}$ 655 nm Tl line. Its shape is very sensitive to charged particles interactions (figure 1) contrary to the 577 nm Hg line previously studied.

We analyse the instantaneous line profiles emitted along a chord of the discharge by using a THR Jobin-Yvon monochromator. Typically we use results at 0 and 5 ms after the arc voltage is zero. Because of modulation of electron density we separate neutral and electron broadening. A typical result is shown on fig. 1. Experiments were performed for several discharges as shown in table 1.

After analysis of broadening mechanisms we compute the local line profile and solve the radiative transfert equation to obtain the apparent line profile. We take into account doppler and pressure broadening. Neutral broadening is mainly due to Hg atoms. For neutral and electron broadening impact approximation is valid and we assume respectively C_6/r^6 an C_4/r^4 interaction potential /1/. For Stark broadening, width 2δ and shift Δ are corrected according to the GRIEM's approximation /2/ to take into account ion effects. The resulting local line profile is the well known VOIGT profile. Temperature profiles and plasma composition have been determined elsewhere /3/ assuming local thermodynamic equilibrium. We take an A_{ij} value equal to $3.9\ 10^6 s^{-1}$ /4/. Broadening parameters C_6 and C_4 are fitted to reproduce experimental profiles for all the discharge conditions studied. The best results are obtained for the following values :

$$C^6(\text{Hg-Tl}) = 45.10^{-42}\ \text{rad.m}^6\text{s}^{-1}\ ;\ C_4 = 145.10^{-22}\ \text{rad.m}^4\text{s}^{-1}$$

Figures 2a and 2b show optimized widths and shifts of the apparent line profile along with experimental ones versus the chord position.

Stark broadening is the most effective mechanism for the 655 nm Tl line. For the 577 nm Hg line the major effect is due to neutral interaction. Work is presently in progress to use simultaneously profiles of both lines for plasma composition determination.

This work is a part of a joint research program between CPAT and University of CRETE and has been supported by Service Recherche de la Compagnie des Lampes et du Matériel d'Eclairage.

/1/ G. TRAVING, Plasma diagnostics, W. LOCHTE-HOLTGREVEN ed., North Holland (1966).

/2/ H.R. GRIEM, Spectral line broadening by plasmas, Ac. Press(1974)

/3/ T. ABDENNADHER, Thesis n°3237, Toulouse (France), (1985).

/4/ D. KARABOURNIOTIS, S.COURIS, J.J.DAMELINCOURT, XVII Int. Conf. on Phen. in Ion. gases, 2, 996, Budapest 8-12 july 1985.

Discharge		Tl2		Tl5	
Instant (ms)		0	5	0	5
Hg pressure (atm)		1.46	1.62	4.35	4.43
T_o(K)		5100	5970	4940	5480
ℓ		3.2	2.2	3	2.6
Tl pressure (torr)		6	3.8	7.5	6.5
Axis densities ($10^{21}m^{-3}$)	e^-	3.17	8.05	2.9	6.25
	Hg^+	0.15	2.03	0.21	1.21
	Tl^+	3.0	6.02	2.7	5.04

Table 1 - Characteristics of the discharges (A.C. power=1030 w). Temperature profiles are given in the form : $T(\rho) = T_o-(T_o-T_w)\rho^{\ell}$

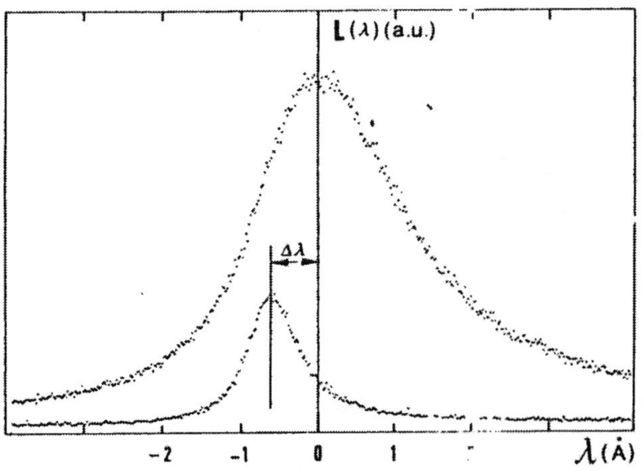

Fig. 1 - Tl 655 nm experimental line profile.

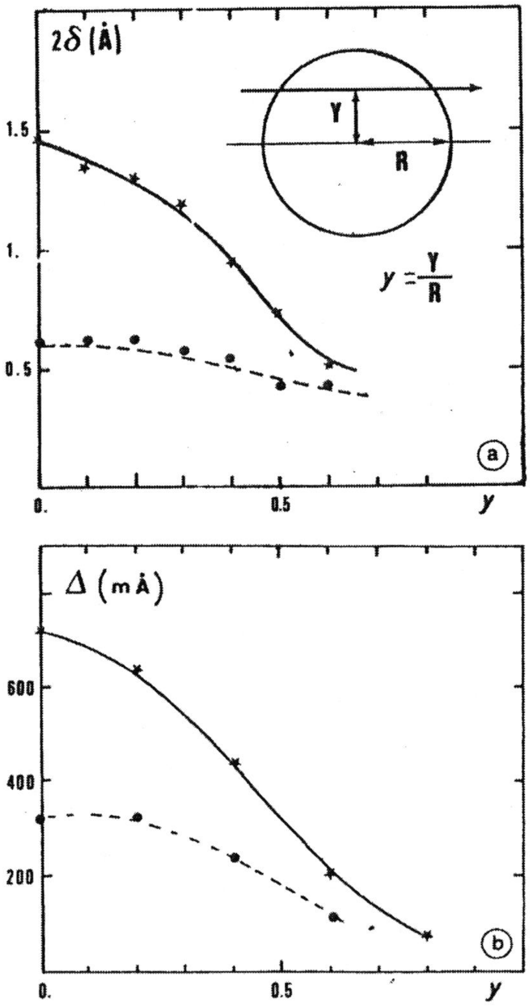

Fig.2-Evolution of width 2δ and shift Δ versus y for Tl2 discharge.

Calc.val.:—(5ms),--(0ms)
Exp. val.: ∗(5ms),● (0ms).

DEVIATIONS FROM THERMOCHEMICAL EQUILIBRIUM IN Hg-TlI LAMP ARCS

D. Karabourniotis, S.Couris, and G.Zissis
Physics Department, University of Crete,71049 Iraklion,Crete,Greece
J.J. Damenlicourt
Centre de Physique Atomique,Laboratoire Associé au CNRS n° 277,
Université P. Sabatier, 31062 Toulouse,France.

In the course of spectroscopic determination of the thallium pressure in Hg-TlI lamp arcs we have observed deviations from thermochemical equilibrium both in the center and in the periphery of the discharge.

The lamps were operated vertically at 100W/cm (5Hz) and time-resolved radiation measuremets have been carried out at the moments of maximum (5 msec) and minimum (0 msec) electron density, for two mercury (~1.4 and 4.7 atm), and different thallium pressures.

We introduce an effective thallium pressure

$$\overline{P}= \int_0^R P(r)f(r)dr / \int_0^R f(r)dr,$$

$$f(r) = \left[Z(T)\ T(r)\right]^{-1} \exp\left[-E_\alpha/kT(r)\right],$$

P(r)= local partial pressure, Z(T)= partition function, T(r)= temperature distribution, and E_α=energy of the thallium metastable level $\alpha = 6^2P_{3/2}$. T(r) and mercury pressure were determined from Abel inversion of the Hg-577 nm line [1]. Assuming LTE the axis partial pressure of thallium ($P_{\alpha x}$) is obtained from Abel inversion of the Tl-655 nm line. Therefore, using the constant-mixing ratio approximation and thermochemical equilibrium,the local pressure has been determined [2], from which the effective thallium pressure ($\overline{P}_{eq}$) is calculated, The effective thallium pressure is also determined experimentally ($\overline{P}_{ex}$) from the characteristics of the self-reversed Tl-535 nm line emitted along a diameter, using the expression [3]

$$P_{ex}= C\ (p\delta/Y), \text{ in which } C= (8\pi^2\nu_o^2/c^2)\ (g_e/A)^{-1},$$

$$Y= \int_0^R f(r)\ dr, \text{ and}$$

p= absorption parameter, δ=line profile half-width.

From Fig. 1 we obtained $\overline{P}_{eq} = A+BP_{ax}$, with A=0.60 Torr, B=1.00± 0.02, and r(correlation coefficient)=0.995.

From Fig. 2 we obtained $\overline{P}_{eq}=A+B\overline{P}_{ex}$, with at o msec: A≃o, B=0.73± 0.04, and r=0.995; at 5 msec: A = 4.35 Torr, B≐0.81±0.10, and r=0.990

Because $\overline{P} = N_\alpha g_\alpha Y$, $N_\alpha = \int_0^R n_\alpha(r)dr$, we found that at o msec $N_{\alpha;ex}/N_{\alpha;eq} =$ 1.37±0.08. This may be explained from overpopulation of the Tl-$6^2\ P_{3/2}$ level due to self-absorption.

At 5 msec from A we found $N_\alpha = 1\times10^{15}\ cm^{-2}$ corresponding to the metastable thallium atoms created by photodissociation of TlI molecules [4], while the slope 1/B = 1.23± 0.15 gives the over-population in the inner region caused by self-absorption.

[1] J.J. Damenlincourt et al, J.Phys.D.11,1029(1978); [2] J.J. Damelincourt et al, J.Phys. D.13,583(1980); [3] D. Karabourniotis et al, J.Appl. Phys. 58,2785(1985; [4] D.C. Fromm et al, Symp.High Temp. Metal Halide Chem. 78-1,61 (1978).

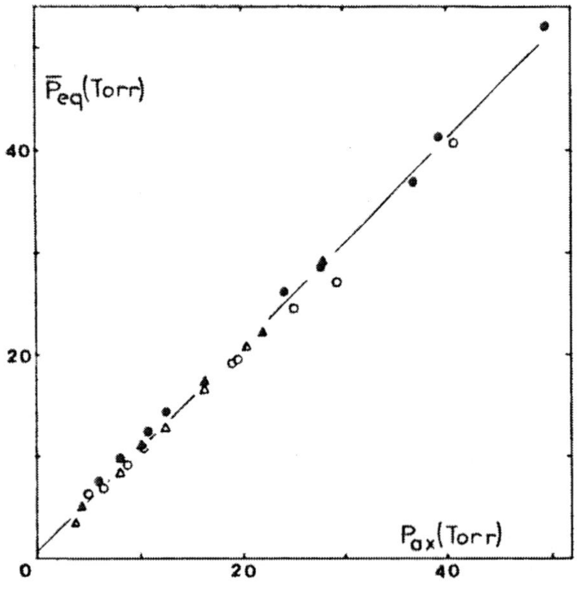

Fig. 1

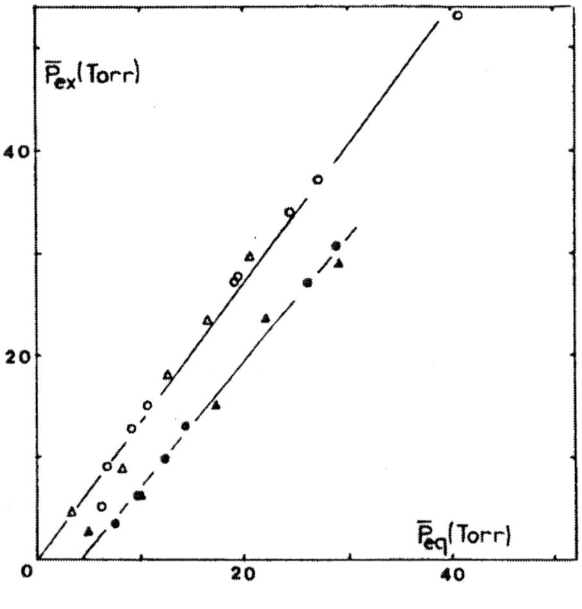

Fig. 2

●: P_{Hg} ≃ 1.4 atm, t=5 msec; o: P_{Hg} ≃ 1.4 atm, t=o msec;
▲: P_{Hg} ≃ 4.7 atm, t=5 msec; Δ: P_{Hg} ≃ 4.7 atm, t=o msec.

"MEASUREMENTS OF THE PHYSICOCHEMICAL PROPERTIES OF LIGHT SOURCES USING NON-LINEAR LASER SPECTROSCOPY"

R. DEVONSHIRE, I.S. DRING, and J. MEADS,
Chemistry Department, The University, Sheffield, S3 7HF, U.K.

H.F. BOYSAN, and J. SWITHENBANK,
Chemical Engineering and Fuel Technology Department, The University, Sheffield, S3 3JD, U.K.

D.A. GREENHALGH,
Engineering Sciences Division, Harwell Laboratory, Oxfordshire, OX11 ORA, U.K.

The potential of non-linear laser spectroscopic techniques for the investigation of operational light sources was discussed at the previous symposium[1], and a more detailed discussion of the scope and limitations of such studies, including a report of some of the results obtained from such investigations by the authors, was presented recently.[2]

This contribution will discuss the application of coherent anti-Stokes Raman scattering (CARS) to the study of incandescent light sources.[3] Temperature contours for a wide range of temperatures have been obtained from gas filled incandescent lamps and the results compared with the predictions of such contours by sophisticated theoretical models of fluid flow. The agreement between the two different approaches was found to be remarkably good.

Small discrepancies between measured and predicted temperature values in certain regions of the recirculating gas flows have been traced to an interesting non-equilibrium effect between the internal and external degrees of freedom of molecules in the flow. A theoretical model of this effect will be presented. We believe the phenomena is the basis of a new experimental technique for the investigation of high temperature gas phase and plasma chemical equilibria and relaxation kinetics, a technique we refer to as recirculating flow relaxation spectrometry (RFRS).

Remarkable CARS spectra obtained from light sources of molecules in thermal equilibrium at extremely high temperatures will also be presented.

References:

1. R. Devonshire. Contributed Paper, Third International Symposium on the Science and Technology of Light Sources, Toulouse, 1983. Abstract No. 38.
2. R. Devonshire. NATO Advanced Study Institute, "Radiative Processes in Discharge Plasmas", Pitlochry, Scotland, July 1985. (To be published).
3. H.F. Boysan, R. Devonshire, I.S. Dring, D.A. Greenhalgh, G.E. Hoey, J. Meads, F.M. Porter, J. Swithenbank, and D.R. Williams (Papers submitted and in Preparation. Preprints available).

A NOVEL INTERSECTING LASER BEAM TECHNIQUE FOR FLUORESCENT LAMP DIAGNOSTICS*

Philip E. Moskowitz
GTE Lighting Products
Danvers, MA 01923

Knowledge of the $Hg(6^3P_1)$ spatial distribution in Hg-rare gas low pressure discharges is important for understanding radiation transport, and aids in the formulation of discharge models(1) for fluorescent lamps. There is a need to perform measurements of this distribution with high accuracy, as small differences in the spatial profile may indicate relatively large effects on the transport of radiation from the center to the walls of the discharge. Considerations important in an experimental effort are: 1) the measurement should not significantly perturb the plasma, and 2) in the case of cylindrical geometries, provision must be made to measure the spatial profile over the entire diameter of the tube, especially near the walls. This last point is due to the fact that photons generated in the bulk of the discharge are absorbed and reemitted 50 to 100 times in their migration towards the wall, and changes in the distribution near the wall could significantly affect the plasma radiant efficiency.

We have been developing a number of diagnostic techniques and assessing their accuracy. We have been using a CW ring dye laser ($\Delta\nu$=1 MHz) for these studies. In evaluating various approaches, we noted that conventional laser absorption techniques utilizing a single beam transverse to the discharge axis require an Abel inversion to extract useful information. Furthermore, measurements in which the beam is parallel to the discharge axis are plagued by edge distortions due to the non-uniformity of the discharge path. Laser Induced Fluorescence (LIF) is an effective technique, but one must avoid problems with altering the collection efficiency of the detector as the laser beam is translated to obtain spatial information. Furthermore, care must be exercised in the interpretation of the signals as one approaches the optically thick regime.

In this paper we report on a novel intersecting two laser beams technique which circumvents many of the problems mentioned above. In addition, it has the unique advantage of giving, for the first time, pinpoint information about the discharge positive column. This experiment makes use of a "pump" and "probe" beam, originating from the same laser, and intersecting at right angles at a small point in the discharge. The laser radiation is at 435.8nm, which couples the Hg (6^3P_1) and Hg(7^3S_1) states. The pump or saturating beam is "chopped" and modulates the Hg(6^3P_1) population along its path. The weak probe samples the modulated density at the intersection zone, providing highly localized Hg(6^3P_1) density information. We will present data for a Hg-Ar discharge containing one isotope (^{200}Hg) of mercury and 2.5 torr of argon.

(1) R. Lagushenko and J. Maya, Journal of IES, p. 306-314 Oct. 1984; ibid Bull. of Am. Phys. Soc. 30, 139 (1985)

* Supported by the US Department of Energy through a subcontract from Lawrence Berkeley Laboratory.

LASER INDUCED FLUORESCENCE STUDIES OF Ba LOSS FROM FLUORESCENT LAMP ELECTRODES WITH CONVENTIONAL BALLASTS

A.K. Bhattacharya and P.G. Hlahol

Lighting Business Group,

General Electric Co., Cleveland, USA.

The operating life of a F-lamp is determined by the life of its electrodes which consist of a W-coil covered with a mixture of Ba, Ca, and Sr oxides. During lamp life the emission material on electrodes gradually disappears as a result of evaporation and sputtering by ion bombardment. The loss of Ba results in a substantial increase in the work function of the cathode leading to lamp failure. The service life of an electrode depends upon the Ba depletion rate which is determined by its operating characteristics. LIF methods for investigating the Ba depletion rate from such electrodes have been developed and are presented. Application of this technique allowed determination of: (1) the heat of sublimation of Ba from a triple oxide coated W-coil, (2) the Ba removal rate in both the anode and cathode half-cycles of an operating lamp, (3) the consumption rate of Ba under various operating conditions, and (4) Ba loss rates from electrodes operating on various electromagnetic ballasts.

A Nd-YAG pumped dye laser was used to generate a 10ns wide pulse of 700.2nm radiation. A frequency doubling angle tuned KDP crystal was used to generate 350.1nm radiation for excitation of Ba in the ground state. The collimated beam of 350.1nm radiation passed through a clear window in the lamp envelope in a plane parallel and approximately 1mm in front of the electrode. The LIF signals at 582.6nm was monitored using a 0.25m monochromator, box-car integrator and a data acquisition system. The Ba emission (LIF at 582.6nm) from an externally heated electrode with the arc off was measured. From such studies the heat of sublimation for the triple oxide-W system was determined to be 60 kcal/mol. The consumption of Ba during steady-state operation of lamps on various ballasts, e.g., linear reactor, standard rapid start, energy efficient, and energy efficient with cathode heat cut-out ballasts, was measured as a function of the phase of the lamp current. The loss rate of Ba depends upon the phase of the lamp current; it is higher in the anode half-cycle than the cathode half-cycle. The increase in Ba emission when the electrode acts as an anode is due to the heating of the electrode hot spot by electron bombardment. The lower rate during the cathode half-cycle can be attributed to thermionic cooling of the hot spot. Thus, the major loss of Ba occurs during the time when the electrode is acting as an anode. Ba depletion is strongly influenced by the operating characteristics of the lamp-ballast system. The Ba loss rate was found to be smallest for the energy efficient ballast and greatest for the rapid start ballast. The loss rate is dependent upon the applied cathode heat and is very sensitive to the lamp current.

The technique discussed here was used for the first time to differentiate Ba depletion rates from an electrode in both cathode and anode half-cycles. It was shown that: (1) the heating of electrodes mainly arises from electron bombardment during the anode half-cycle, (2) Ba loss is mostly due to thermal evaporation from the electrode hot spot, and (3) LIF can be used for studying the evaporation processes of lamp electrodes.

MAPPING OF MERCURY AND XENON DENSITIES IN HIGH PRESSURE LAMPS

B. Weber

Philips GmbH Forschungslaboratorium Aachen, Germany

Weisshausstr. , 5100 Aachen

Absorption measurements with a UV laser can yield spatially resolved values of mercury and xenon densities even in the cold regions of the lamp. This information is necessary for reliable calculations of conductivity and radiative transport in high pressure lamps.

The method is based on the measurement of the absorption coefficient in the far wing of the 253.7 nm resonance line of Hg at 257.2 nm /1/. The latter is the wavelength of the frequency doubled Ar ion laser. Due to the fact, that the 253.7 nm line is broadened by binary interaction among the mercury atoms and between mercury and xenon, the absorption coefficient can be expressed by:

$$k(\ 257.2\ \text{nm}\ ) = A\ n_{Hg}^2 + B\ n_{Hg}\ n_{Xe}$$

The constants A and B have been determined in oven experiments using cuvettes filled with different mixtures of Hg and Xe.

The applicability of this method has been tested on two cylindrical lamps, one filled with Hg and a negligible amount of Xe, the other with Hg and 600 Torr (cold) Xe. The absorption measurements have been performed with an intra cavity doubled Ar ion laser with an UV output power of 80 mW. Lock-in technique and a monochromator have been used to separate the laser light from the discharge emission. An Abel inversion has provided the local absorption coefficients, from which the local paricle densities could be derived using (1). The total amount of Hg in the lamp has been compared to the filling charge of the lamp.

/1/ R. Schäfer, H.-P. Stormberg, submitted to Appl. Phys.

Total amount of Hg in the lamp.
filling: 20mg

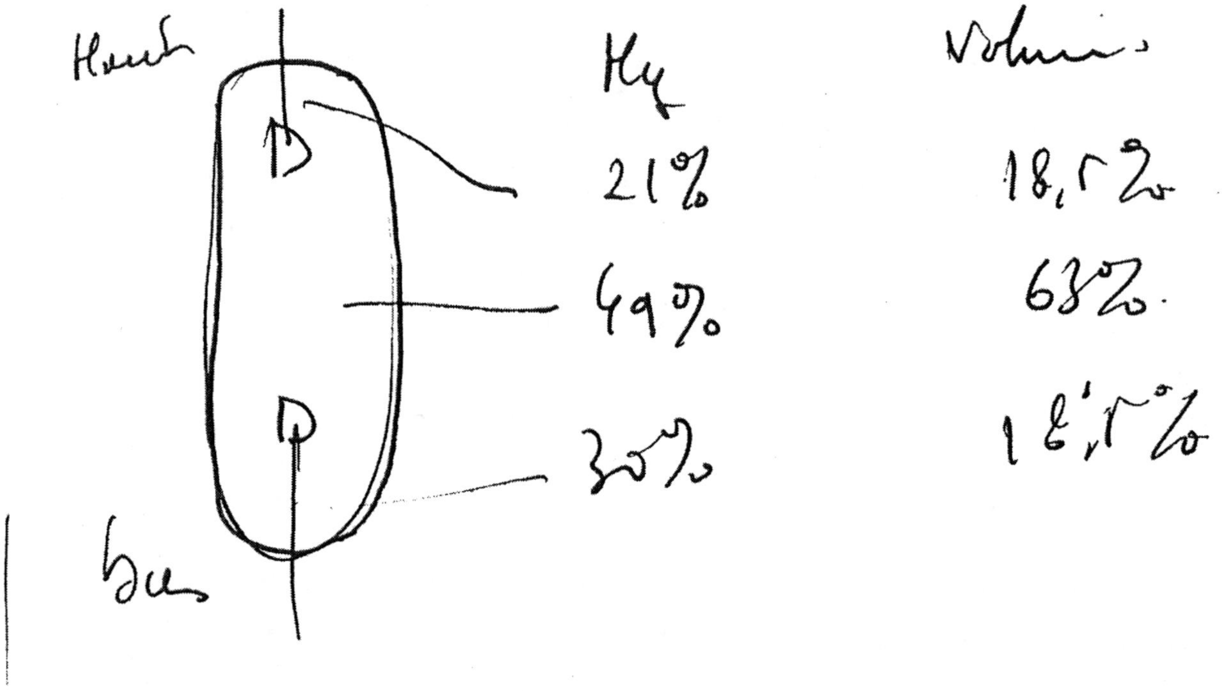

Gap in pagination due to formatting issues.

Pages 227-228

LAMP CHEMISTRY

63 - 69

THE DERIVATION OF THERMOCHEMICAL PARAMETERS FOR USE IN MODELLING METAL HALIDE LAMPS

S A Mucklejohn and N W O O'Brien, THORN EMI Lighting Limited, Research and Engineering Division, Melton Road, Leicester, LE4 7PD, U.K.

The calculation of the chemical composition in the plasma region is a prerequisite for the development of a model for high-pressure discharge lamps. This paper illustrates, using our work on tin(II) chloride, how the relevant thermochemical functions are derived.

For our database, the required thermochemical function is the temperature variation of the Gibbs free energy for the formation of the species from the corresponding gaseous elements, equation (1).

$$M(g) + nX_2(g) = MX_{2n}(g) \tag{1}$$

This is expressed as:

$$\Delta_f' G_m^\circ(T)/\text{kcal.mol}^{-1} = A + Bt + Ct^2 + Dt\ln t \tag{2}$$

where $t = T/1000$ K and $T = 298$ to 6000 K.

It is important to note in equation (1) that the reference state of the elements, at all temperatures, is taken to be the gas phase.

Tin halides are not included in the JANAF tables; the compilation by Barin, Knacke and Kubaschewski (1) contains some information on tin species, but the temperature range is limited.

For the reaction under consideration in this work, equation (3), the Gibbs free energy change is given by equation (4).

$$Sn(g) + Cl_2(g) = SnCl_2(g) \tag{3}$$

$$\Delta G_m^\circ(T) = G_m^\circ(SnCl_2,g,T) - [G_m^\circ(Sn,g,T) + G_m^\circ(Cl_2,g,T)] \tag{4}$$

The more usual form of equation (4) is that given in the JANAF tables, equation (5).

$$\Delta_f G_m^\circ(T) = \Delta_f H_m^\circ(T) - T[S_m^\circ(P,T) - S_m^\circ(R,T)] \tag{5}$$

where:

$$\Delta_f H_m^\circ(T) = \Delta_f H_m^\circ(298\ K) + [H_m^\circ(P,T) - H_m^\circ(P,298\ K)] - [H_m^\circ(R,T) - H_m^\circ(R,298\ K)] \tag{6}$$

and P = product $\{(SnCl_2(g)\}$; R = reactants $\{Sn(g), Cl_2(g)\}$.

The enthalpy of formation of $SnCl_2$(s), $\Delta_f H^\circ_m(SnCl_2,s,298\ K)$, is available from the literature and can be transformed into that for the gas by inclusion of the enthalpy of sublimation (or, alternatively, the enthalpies of fusion and evaporation).

Thermochemical functions for Cl_2(g) and Sn(g) are available from the literature and will not be considered further here.

The functions $S^\circ_m(SnCl_2,g,T)$ and $[H^\circ_m(SnCl_2,g,T) - H^\circ_m(SnCl_2,g,298\ K)]$ are derived from $S^\circ_m(SnCl_2,g,298\ K)$ and the molar heat capacity, $C^\circ_{p,m}(SnCl_2,g,T)$. These quantities are calculated from spectroscopic constants available from the compilation by Krasnov (2).

The enthalpies of vaporization of $SnCl_2$ are derived from a detailed study of the vapour phase above the condensed phases. We have employed three methods in this work: Knudsen effusion/mass spectrometry; vapour pressure measurement; the transpiration technique.

The first gives the species present in the vapour ($SnCl_2$,Sn_2Cl_4 and a small amount of $SnCl_4$); the second gives the total vapour pressure over a wide temperature range and the third yields the monomer:dimer ratio. Information from the first two techniques is essential for the latter. From these results the partial pressures of $SnCl_2$, Sn_2Cl_4 and $SnCl_4$ are calculated.

The partial pressures are used to determine the enthalpy of evaporation by the second-law method (from the slope of a plot of $\ln(p/p^\circ)$ against reciprocal temperature) and by the third-law method (using thermochemical functions derived from $S^\circ_m(SnCl_2,g,298\ K)$, $C^\circ_{p,m}(SnCl_2,g,T)$ and $C^\circ_{p,m}(SnCl_2,l,T)$. If the results from the second-law and third-law methods are in good agreement, we consider our data to be reliable and we are confident to use the parameters $\{S^\circ_m(SnCl_2,g,298\ K)$, $C^\circ_{p,m}(SnCl_2,g,T)$ and $\Delta_{evap}H^\circ_m(SnCl_2,298\ K)\}$ in the calculation of $\Delta_f G^\circ_m(SnCl_2,g,T)$.

REFERENCES

1. "Thermochemical Properties of Inorganic Substances", I Barin, O Knacke and O Kubaschewski, Springer-Verlag, Berlin (1977).
2. "Molecular constants of inorganic compounds", (in Russian) Khimiya (1968).

PRELIMINARY REPORT ON THE EXAMINATION OF ELECTRODE EROSION BY THE TRANSPIRATION TECHNIQUE.

N W O'Brien and S A Mucklejohn, THORN EMI Lighting Limited, Research & Engineering Division, Melton Road, Leicester, LE4 7PD, U.K.

The phenomenon of electrode erosion has been recognised since the early investigations on the addition of various metal halides to high-pressure mercury discharges. We define electrode erosion simply as the removal of material from any part of the electrode. Although the consequences of electrode erosion are well established (wall blacking, rise in operating voltage and limited lamp life), the mechanisms of such erosion are not well understood. Tungsten is removed from various regions of the electrode, since these regions are not at the same temperature, several different processes must be occuring. The reactions are thought to involve free halogens (X and X_2) and oxygen. The former are assumed to remove tungsten by the formation of $WX_n(g)$ (n = 2,4,5). The combination of oxygen and halogens are assumed to react to form $WO_nX_{4-n}(g)$ (n = 1 or 2).

The most detailed consideration of the transport cycles within a high-pressure discharge lamp has been given in two papers by Speros *et al.*[1,2] In an elegant series of experiments they showed the transport of tungsten was governed by the amount of oxygen and halogen available at, or close to, the electrode. All experiments were carried out in silica bodied lamps.

In an attempt to develop a more detailed understanding of the chemical processes involved in electrode erosion, we have designed experiments to study the transport of tungsten in the presence of metal halide vapours to which controlled amounts of impurities are added. The reactions of some $W(s)/0_2(g)/X_2(g)$ systems have been studied by several authors under equilibrium conditions but the properties of $W(s)/0_2(g)/MX_2(g)$ systems have not.

Our system is based on the transpiration technique, whereby a flow of inert gas is allowed to pass over a solid or liquid sample at a rate such that equilibrium can be established between the carrier gas and the vapour above the sample. The transported vapour is condensed and subsequently analysed. If the total vapour pressure and the vapour phase species are known, the saturated vapour composition can be determined.

In our system a resistively heated tungsten filament (held at constant current) is positioned above a metal halide sample within the transpiration apparatus. The gas flow through the apparatus can be either pure argon or argon together with an added impurity (e.g. O_2,H_2,H_2O,CO,CO_2). The first case monitors the reactivity of the metal halide vapour towards the tungsten and the latter monitors the combined reactivity of the metal halide and the selected impurity. Gas flows are accurately governed by electronic mass flow controllers. By choice of suitable gas mixtures and flow rates the oxygen concentration in the carrier gas can be varied from a minimum of 2 ppm(V) to >5%.

The amount of tungsten transported and condensed in the collection tube can be quantified by newly developed analytical methods. The inherent problems of measuring sub-milligram quantities of tungsten are exacerbated by the metal halide matrix.

The aim is not to simulate conditions found within a discharge lamp but to study the relevant chemical reactions in a controlled, reproducible environment. Thus the effect of the selected impurities can be monitored individually.

The erosion of the tungsten filament can be studied over a wide range of conditions: by varying the temperature (and hence temperature profile) of the filament; by varying the temperature and nature of the metal halide sample; by varying the concentration and nature of the minority component of the carrier gas.

Our experiments have been supplemented by observations of a series of lamps containing various metal halide systems and by chemical composition calculations.

REFERENCES

1. D M Speros, R M Caldwell and W E Smyser, High Temperature Science, 4, 99 (1972).
2. D M Speros, R K Datta and D A Lynch, J. Electrochem. Soc., 128, 778 (1981).

THE REACTION OF RARE-EARTH METAL IODIDES WITH THE SILICA WALL IN METAL HALIDE DISCHARGE LAMPS.

W. van Erk and T. Rietveld
Philips Lighting Division
P.O.box 218, 5600 MD Eindhoven, the Netherlands.

Chemical processes occuring in metal halide lamps are very complex. Extra complications arise when rare-earth halides are present in the discharge tube. The phenomena which can be observed in these types of lamps are transport of silica from hot areas of the wall to colder places and transport of silicon to the electrode.

In both transport processes the silicon iodides SiI_4 and SiI_2 play an important role. They are formed by the reaction of the rare-earth iodides with the fused silica wall at the high temperatures at which these lamps are operated. In case of thulium iodide thermodynamic calculations predict the formation of thulium oxyiodide (a) and not of thulium oxide.

$$SiO_2 + 2REI_3 \rightleftharpoons 2REOI + SiI_4 \qquad (a)$$

However, in aged lamps not the oxyiodide is found but thulium silicate:

$$7SiO_2 + 4REI_3 \rightleftharpoons 2RE_2O_3.2SiO_2 + 3SiI_4 \qquad (b)$$

Unfortunately reliable thermodynamic data of thulium silicate are not yet available.

The gaseous silicon iodides formed at the silica wall will decompose near the hot electrode and elementary silicon is formed. The silicon will dissolve in the tungsten electrode. The net reaction is: $2Hg + SiO_2 + 2REI_3 \rightarrow 2REOI + 2HgI_2 + Si \text{ in } W \qquad (c)$

Due to this process the iodine pressure will increase during life.

Thermodynamic calculations show that the transport of silica is a chemical transport process. Two mechanisms are possible:

$$\left.\begin{array}{l} SiO_2 + SiI_2 \rightleftharpoons 2SiOI \\ SiO_2 + SiI_4 \rightleftharpoons 2SiOI_2 \end{array}\right\} \quad (d)$$

$$SiO_2 + SiI_4 \rightleftharpoons 2SiO + 4I \qquad (e)$$

According to our calculations the vapour pressures of the oxyiodides are higher than the pressure of SiO indicating that mechanism (d) is the dominant chemical transport process. However, the existence of the silicon oxyiodides has not been proven yet. Their thermodynamic data were estimated using the data of other oxyhalides.

The silicon iodides are formed by reaction (a) or (b). The pressure of SiI_4 will therefore depend on:

1) The temperature of the molten salt mixture of REI_3 and other additives.
2) The concentration of REI_3 in this mixture.
3) The rare-earth which is used.

In order to verify this model experimentally 150W lamps were made filled with NaI/ScI_3, NaI/TmI_3 and NaI/DyI_3 with the molar ratio $NaI/REI_3 = 2/1$ and for NaI/ScI_3 also with the ratio 21/1. These lamps were burned in a horizontal position at a wattage of 150W. The intensities of the optically thin lines of silicon, iodine and mercury were measured as well as the $\Delta\lambda$'s of the self-reversed emission lines of sodium and of the rare-earth metal. The results of this study will be presented and will be compared with the results of thermodynamic calculations. It will be shown that these type of calculations are a powerfull tool in the understanding of the complex chemical processes occuring in metal halide lamps.

COMPLEX FORMATION IN METAL HALIDE LAMPS. SODIUM ENHANCEMENT FOR THE SYSTEMS NaI + SnI_2 AND NaI + PbI_2

J. Seehawer and K. Hilpert*
OSRAM GmbH, D-8000 München 90
* Institut für angew. Phys. Chemie, KFA, P.F. 1913, D-5170 Jülich

The operation of metal halide lamps of low color temperature requires a relatively high sodium pressure in the arc core. In order to enhance the sodium constituent the additional dosing of tin halide has turned out to be very effective due to formation of a sodium-tinhalide complex [1,2,3]. Moreover, there is evidence that also a stable sodium-lead complex exists [4]. The aim of this work was to investigate the equimolar systems NaI + SnI_2 and NaI + PbI_2 by measuring spectroscopically the various vapor pressures in discharge vessels which were operated with their lower end in a tin bath, adjusting in this way the cold-spot-temperature.

Experimental: The vessels (electrode distance 2.5 cm, inner diameter 0.8 cm), filled with Ar (40 mb), Hg (60 mg), and 1) NaI, 2) NaI + SnI_2 (aequimolar), 3) NaI + PbI_2 (aequimolar) were operated vertically at 175 W, 50 Hz. Time resolved spectral radiance measurements were conducted in a horizontal midplane of the arc. In order to keep the ionization and segregation effects as low as possible the measurements were preferably carried out at the moments of minimum intensity. The evaluation procedure was the same as outlined in [5]. The method is based on the approximation of a parabolic temperature profile and a constant partial pressure within the arc core. The arc temperature was derived, avoiding absolute calibration, by comparing the peak intensity of the self-reversed optically thick line Hg 5461 Å and the optically thin line Hg 5770 Å (transition probality $g \cdot A = 1.7 \cdot 10^8\ s^{-1}$[6]).
On the other hand this line has been used as a comparative standard for the determination of the Na-, Sn-, and Pb-pressure utilizing the following transitions: Na 5683/5688 Å, Na 6154/6161 Å, Sn 5632 Å, Pb 7229 Å.

All measured partial pressures were corrected for ionization. In order to check the spectroscopic measurements fundamental mass spectrometric investigations were carried out on the systems NaI, SnI_2, PbI_2, NaI/SnI_2, and NaI/PbI_2.

Results:

1) NaI: Within the scattering range (about a factor of 2) the "constituent-pressure" $P_{Na} + P_{Na^+}$ determined in the arc center is in accordance with the mass spectrometric results [1,7] for $P_{Na} + 2\ P_{Na_2I_2}$.

2) NaI + PbI_2: No significant sodium-enhancement could be observed in the arc. Accordingly the massspectrometrically derived enhancement factor (1.4 at 1000 K) is relatively low.

3) NaI + SnI_2: A high sodium-enhancement could be determined in the arc core. This result could be verified by the mass-spectrometric measurements (Enhancement factor = 62 at 800 K, and = 11 at 1000 K).

References:

[1] A detailed publication of this work will appear in early 1986: K. Hilpert, J. Seehawer: Techn.-wissensch. Abhandl. der OSRAM-Gesellschaft 12, 31 (86)

[2] D.C.Fromm, J. Seehawer, W.-J. Wagner: Light. Res. and Techn. 8 136 (76)

[3] K. Hilpert, K.A.Gingerich: Advances in Mass Spectrometry 9, Proc. 9th Int. Mass Spectrometry Conf., Wien (Aug./Sept.82)

[4] K. Hilpert, H. Gerads, D. Kobertz, R. Viswanathan: 3rd Intern. Symp. on the Science and Technology of Light Sources, Toulouse (Apr. 1983), p. 152

[5] D.C. Fromm, J. Seehawer, W.-J. Wagner: "Proc. of the symposium on high temperature metal halide chemistry", Ed. by D.L. Hildenbrand and D.D.Cubicciotti. Electrochem. Soc. Proceedings Vol. 78-1, p. 61 (1978)

[6] P. van de Weijer, R.M.M. Cremers: J. Appl. Phys. 54, 2835 (83)

[7] K. Hilpert: Ber. Bunsenges. Phys. Chem. 88, 132 (84)

THERMODYNAMIC STUDIES FOR METAL HALIDE LAMPS

K. Hilpert
Institute of Applied Physical Chemistry, Nuclear Research Centre (KFA), P.O. Box 1913, D-5170 Jülich, Germany

J. Seehawer
OSRAM Research, P.O. Box 900620, D-8000 München, Germany

B. Saha
Reactor Research Centre, Kalpakkam 603 102, Tamil Nadu, India

Various metal halide lamps have been developed for different applications in recent years. Lamps with high efficiency for the light generation, a long life and a good colour rendition were obtained. In spite of this success there is still a great potential for the development of metal halide lamps. In order to exhaust this potential, there is an urgent demand for systematic investigations on the vaporization of metal halides and on the thermochemistry of their vapours. Such investigations were carried out by us on different systems (1) elucidating important aspects of lamp chemistry which are of key importance for the practical application. Some of the results obtained from our recent investigations on the system NaI-DyI_3 will be reported. The system NaI-DyI_3 is for example important for lamps of the types HQI-NDL and HQI-D manufactured by OSRAM.

The investigations were carried out using Knudsen effusion mass spectrometry. Samples with different compositions x mol% NaI / (1-x) mol% DyI_3 (x = 0.1, 0.2, 0.3 ... 0.9) were prepared by melting together different mixtures of the component metal halides in sealed quartz ampoules. The vaporization of these samples was studied in the temperature range 970 to 1050 K. The molecules NaI(g), $(NaI)_2$(g), DyI_3(g), $(DyI_3)_2$(g), $(DyI_3)_3$(g), $NaDyI_4$(g), and Na_2DyI_5(g) were identified in the equilibrium vapour over the different samples. Their partial pressures were determined with the exception of $(DyI_3)_2$(g) and $(DyI_3)_3$(g). Isotherms for the partial pressures of these species were evaluated at 1000, 1050 and 1100 K as function of the composition over the complete range of the NaI-DyI_3 system.

The partial pressures of all gaseous species remained constant at temperatures of 1000, 1050 and 1100 K for molar NaI/DyI_3 ratios in the condensed material smaller than 0.5 ± 0.1. The pressures of DyI_3(g) as well as $NaDyI_4$(g) decrease and those of NaI(g) as well as $(NaI)_2$(g) increase if the NaI/DyI_3 ratio is enlarged on going from 0.5 ± 0.1 up to 1. This behaviour of the partial pressures agrees with the phase diagram and Raoult's law. The pressure of Na_2DyI_5(g) is by more than two orders of magnitude smaller than that of $NaDyI_3$(g). Dissociation enthalpies of the geaseous homoeo and hetero complexes observed were evaluated.

The following points are particularly important for practical applications in lamps:

1. Dy is enhanced at temperatures of 1000, 1050 and 1100 K by formation of the hetero complex $NaDyI_4$(g).
2. The pressure isotherms measured are an important basis to select an optimal composition of the lamp filling.

References

1) $NaI-SnI_2$, $NaI-PbI_2$, NaI, SnI_2, PbI_2 in: K. Hilpert and J. Seehawer, Tech. wiss. Abh. OSRAM Ges., in press; NaI in: K. Hilpert, Ber. Bunsenges. Phys. Chem. 88, 132 (1984); FeI_2 in: K. Hilpert, R. Viswanathan, K.A. Gingerich, H. Gerads, and D. Kobertz, J. Chem. Thermodyn. 17, 423 (1985); SnI_2, PbI_2, ScI_3 in: Ber. Bunsenges. Phys. Chem. 89, 1292 (1985); ZnI_2 in: J. Chem. Phys. 83, 5227 (1985); $NaI-SnI_2$ in: K. Hilpert and K.A. Gingerich, Int. J. Mass Spectrom. Ion Phys. 47, 247 (1983).

MASS SPECTROMETRIC STUDIES OF THE ATTACK OF TUNGSTEN BY BROMINE, CHLORINE AND OXYGEN

A. Popovič, J. Marsel and O. Kaposi*

J. Stefan Institute and Department of Chemistry, University of Ljubljana, 61001 Ljubljana, Yugoslavia

*L. Eötvös University, Department of Physical Chemistry, H-1088 Budapest, Puskin u. 11-13, Hungary

The use of tungsten oxyhalide and halide cycles in halogen filled incandescent lamps has contributed to the current interest in high temperature investigations of these systems.

In our recent investigations a double compartment Knudsen-cell mass spectrometer assembly was developed for studying the oxydation kinetics of NiTe [1] and the thermal desorption of bromine from polycrystalline tungsten surface [2]. The similar experimental technique was used in the present work.

The pressure and temperature dependence of reaction rate of polycrystalline tungsten with bromine, chlorine and oxygen was determined and the results were presented as Arrhenius plots (Fig.1). From the Arrhenius plots the following brief conclusions can be made:

1.) The good agreement of our experimental data on the reaction rate for the tungsten oxygen reaction with those of other authors using different methods is a strong proof of our experimental procedure.
2.) The reaction of bromine and chlorine with an oxidized tungsten surface is faster than the reaction with bare surface. Furthermore, the reaction of bromine and chlorine with an oxidized tungsten surface proceeds at significantly lower temperatures than the reaction with pure tungsten.
3.) It can be seen from Fig. 1 that chlorine reacts much more faster with tungsten than bromine at the same temperature.

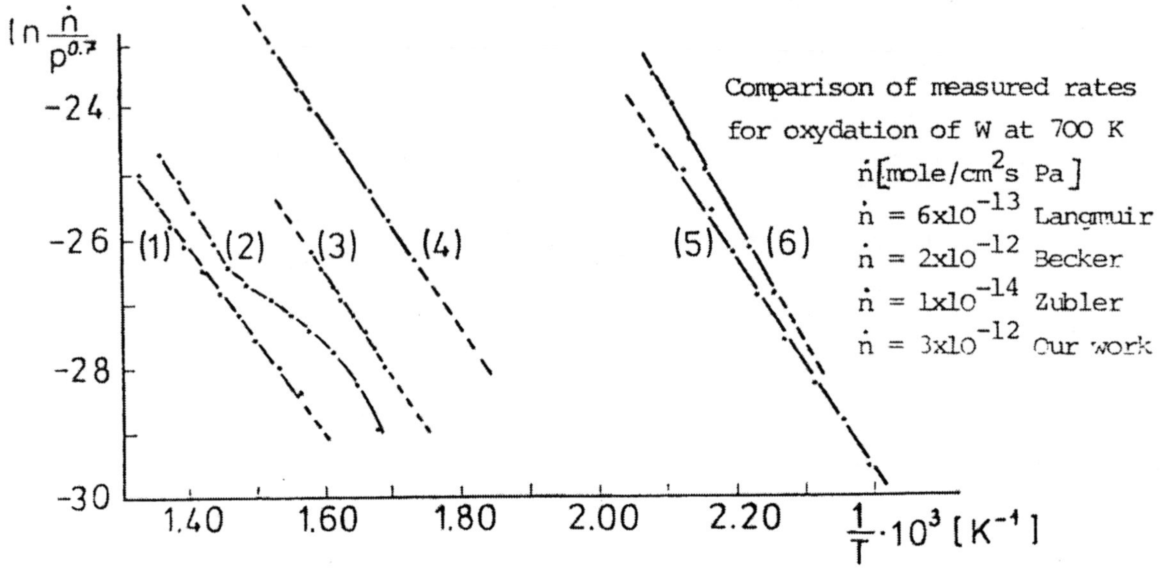

Fig. 1. Arrhenius plots for reactions of polycrystalline tungsten with bromine, chlorine and oxygen.

1) $W+O_2$; $\dot{n}=2.5\times10^{-3}\ p^{0.7}\ \exp(-119\ kJ/RT)$
2) $W+(Br_2+10\%O_2)$
3) $W+Br_2$; $\dot{n}=3.7\times p^{0.7}\ \exp(-144\ kJ/RT)$
4) $W+Cl_2$; $\dot{n}=0.9\times p^{0.55}\exp(-124\ kJ/RT)$
5) $W_xO_y+Br_2$; $\dot{n}=4.1\times10^{3}\ p^{0.7}\ \exp(-131\ kJ/RT)$
6) $W_xO_y+Cl_2$; $\dot{n}=8.44\times10^{7}\ p^{0.55}\exp(-166\ kJ/RT)$

4.) The reaction rate of bromine with bare tungsten wool calculated from the Arrhenius plot at 600 K is 9×10^{-13} $mol\cdot cm^{-2}s^{-1}Pa^{-1}$ shows excellent agreement with the result calculated by the semi-theoretical rate equation used by Rouweler et al. [3] ($\dot{n}=4.5\times10^{-12}$ $mol\cdot s^{-1}cm^{-2}Pa^{-1}$).

1.) A. Popovič and J. Marsel, Adv. Mass. Spec. Vol. 8A. Ed. A.Quayle, pp. 402-407, Heyden, London, 1980.

2.) O. Kaposi, A. Popovič and J. Marsel, Proceedings of the Symposium on Science and Technology of High Temperature Light Sources. Proc. Vol. 85-2. pp. 78-94. Ed. E.G.Zubler, Toronto, 1985.

3.) G. Rouweler and B.J. de Maagt, Philips J. Res. 33, 1-19 (1978)

SODIUM LOSS IN HPS-LAMPS AND ITS INVESTIGATION BY RADIOCHEMICAL METHODS

K. Weinmüller
OSRAM GmbH, D-8000 München 90

Low-power high-pressure sodium lamps (HPS-lamps) show during lamp life an increasing sodium-loss of the amalgam-filling, which shortens the life of these lamp types, due to increasing lamp voltage.

For investigation of the sodium loss mechanism two very sensitive radiochemical methods were applied:

- neutron activation analysis
- radioactive tracer method with Na-22

1. Neutron activation analysis

Neutron activation analysis was applied to determine the sodium content in the Al_2O_3-ceramic of the arc tubes. The samples were irradiated with thermal neutrons in nuclear reactor in order to activate the sodium by the nuclear reaction Na-23 (n,γ) Na-24.

The radioactive Na-24-isotop decays with a halflife of 15 hours and can be easily detected by its characteristic γ-lines with a Ge(Li)-semiconductor-detector. The intensities of these γ-lines are proportional to the Na-content in the sample.

This analytical method however is not applicable for the electrode systems and the outer glass-bulb.

2. Na-22 - tracer method

The analytical limitation by disturbing activities can be avoided by application of the radiochemical tracer method with the nuclide Na-22.

For this purpose the Na-amalgam of the lamp-filling was marked with the radionuclide Na-22, which can be detected by its characteristic γ-lines. Although the detection-limit for this method is somewhat higher than for neutron activation analysis, it offers the possibility of analysing Na of the amalgam-filling also on the electrode systems or on the outer glass-jacket of the lamp.

The combination of both analytical methods allowed to locate and to quantify the sodium contents in different parts of the arc tube, glass-sealings, electrode systems and outer bulb and to determine the sodium loss rate depending on the burning time.

According to the results obtained so far, diffusion processes through the hottest area in the middle section of the arc tube wall and into the glass-sealings are responsible for the sodium loss of low-power high-pressure sodium lamps.

ELECTR.BALLASTS

70 - 74

INVESTIGATION INTO THE EFFECT OF THE CURRENT WAVEFORM MODULATION ON FLUORESCENT LAMPS AT HIGH FREQUENCY

Richard J. Citino
North American Philips Lighting Corporation
330 Lynnway
Lynn, MA

It is generally known within the lighting industry that there are advantages to operating fluorescent lamps at high frequency (≈ 25 KHz) relative to mains frequencies of 50 or 60 Hz [1-5]. For example, in a non-modulated high frequency system, the lamp power factor approaches 1, the positive column efficiency is increased at a reduced current, and the anode fall is eliminated resulting in a significant increase in lamp efficacy.

A major factor affecting lamp performance addressed in this paper is the amount of amplitude modulation or amplitude distortion that is introduced into the high frequency lamp current waveform.

In order to convert mains frequencies (50 or 60 Hz) to high frequency (≈ 25 KHz), electronic ballasts rectify the A.C. input to produce D.C. as shown in Figure 1-A and 1-B. Figure 1-C illustrates how filtering will affect the D.C. signal. Figure 1-D shows the resultant output waveforms of a high frequency inverter/amplifier when driven by filtered and non-filtered sources. When the HF voltage waveforms of Figure 1-D are applied to fluorescent lamps, they produce current waveforms through the lamps as illustrated in Figure 1-E.

Referring to the current waveforms of Figure 1-E, it can be seen that the top current waveform is essentially zero for a period period T which reduces lamp efficacy due to the lamp cycling on and off. The bottom current waveform of Figure 1-E however, shows that the lamp is on continuously thereby improving its efficacy. Further improvements in lamp efficacy can be achieved by minimizing or eliminating the amplitude distortion of lamp current waveform.

Preliminary results on a standard F40T12 cool white fluorescent lamp show an approximate 7% gain in lamp efficacy when a lamp is operated on a non-modulated high frequency system vs. a "100% modulated" high frequency system.

Over the past several years, commercial electronic ballasts have become available which offer improved system efficiencies relative to conventional ballast systems. However, electronic ballasts may not operate fluorescent lamps at their optimum level, due to the reason noted in the preceding paragraph.

Results from an in-depth analysis into the effects of "current waveform modulation" on some of the most common fluorescent lamp types will be presented.

[1]Edward E. Hammer, "Fluorescent Lamp Operating Characteristics at High Frequency" Journal of I.E.S. Volume 14/Number 1, October, 1984, pp. 211-222.

[2]Dale E. Hitchcock, "High Frequency Characteristics of 32 Watt T8 Lamps" Journal of I.E.S. Volume 13/Number 1, October, 1983, pp. 26-29.

[3]John W. Waymouth, "Electronic Discharge Lamps" M.I.T. Press, Cambridge, 1971, p. 110.

[4]W. Elenbass, "Fluorescent Lamps" Philips Technical Library, Second Edition, 1971, pp. 170-172.

[5]Edward E. Hammer, Cais A. Ferreira, "F40 Fluorescent Lamp Considerations for Operation at High Frequency" Paper presented at the Annual Conference of the I.E.S. Detroit, MI - 1985.

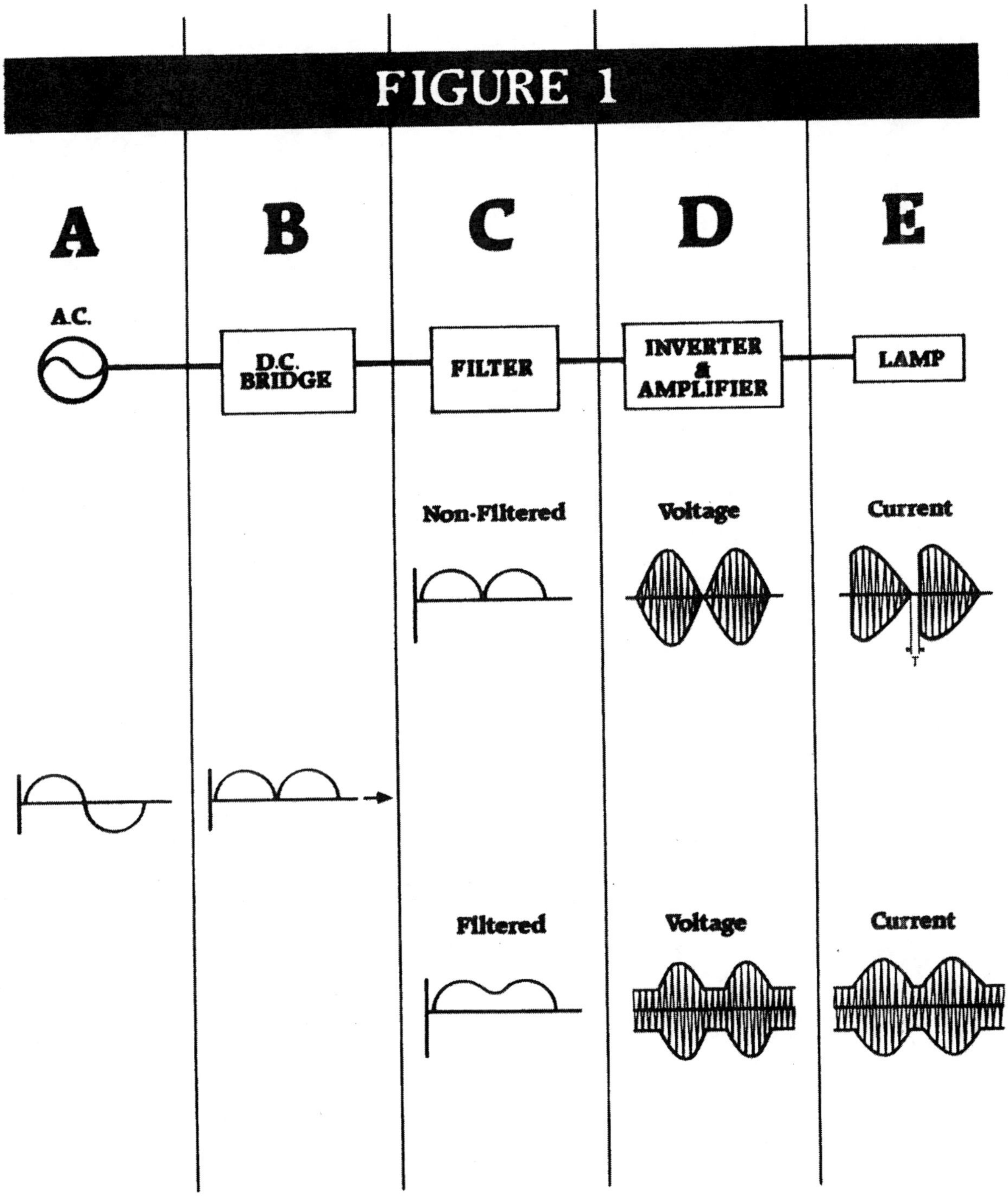
FIGURE 1
A
B
C
D
E
A.C.
D.C. BRIDGE
FILTER
INVERTER & AMPLIFIER
LAMP
Non-Filtered
Voltage
Current
T
Filtered
Voltage
Current

An autotracking system for stable hf operation of HID lamps

F. Bernitz

OSRAM GmbH
D-8000 München 90

Economic operation of modern low and medium power HID-lamps requires the development of electronic controlled gear (ECG). These ballasts have to be designed according to the following aspects: small dimensions, minimum power losses, high efficacy and constant emission of light. The selection of a suitable operating frequency is of great importance. Tests on HQI-TS 70 W lamps showed instable operating conditions caused by acoustic resonances between 8 and 20 kHz and between 25 and 200 kHz as well.

Operating frequencies in the range of about 500 Hz to 18 kHz are eliminated because of the ECG acoustic noise level. The lamp luminous efficacy decreases above 200 kHz, but a markable increase (3...10 %) in comparison to the efficacy at line frequency operation was registrated between 2 and 25 kHz. Therefore we got the best operating frequency between 20 and 25 kHz. However for stable operation, sweeping of the operating frequency is indispensible in this range to avoid some possible instabilities. Because of the tolerances in series production and ageing of the lamps it is not always possible to use a fixed center frequency or a fixed sweep frequency for all lamps. Furthermore each lamp should be operated stable at each ECG.

In this paper a possible solution is presented using an intelligent electronic system in our application called "autotracking system". The idea is to control one of the stability influencing parameters of the operating system, in cases of instabilities.

Such an autotracking system offers 3 basic possibilities to reach stable operation of the lamps changing at least one of the following parameters:

a) center frequency
b) sweep frequency
c) sweep frequency span

Therefore a sensor is required to recognize arc instabilities. Tests showed that an evaluation of lamp voltage or lamp current is an excellent way to get the necessary information. After filtering and processing it is possible to get a signal able to control the autotracking system.
Changing the center frequency is only possible on a small scale because it causes variations of the lamp power. These variations can only be avoided by adjusting the DC-supply voltage or changing the current limiting inductance automatically. For that reason it is easier to change the sweep frequency or sweep frequency span because this has no influence on the lamp power, but also a pronounced influence on lamp stability.

Adequate electronic circuits will be presented.

The result of a reasonable amount of tests with lamps HQI-TS 70 W/WDL is that stable operation can be reached for all lamps by using a fixed center frequency and changing only the sweep frequency between 100 Hz and some kHz.

Realization of electronic ballasts for HID-Lamps

H.-J. Fähnrich

OSRAM GmbH
D-8000 München 90

Electronic control gear (ECG) for operation of High Intensity Discharge Lamps must be very advantageous to pay for itself. What are the attributes of an ECG?

Attribute	Operation system with	
	high frequency	line frequency
Weight	+ +	- -
Size	o	o
Constant wattage	+ +	- -
System efficacy	+	-
Ignition of cold lamps	+ +	-
Ignition of hot lamps	+ +	- -
50 Hz flicker	+	-
Acoustic resonances	- -	+ +

The chart shows that the ECG has nearly only advantages over the magnetic ballast, detrimental are only the acoustic resonances. These resonances being dealt with in several papers must be avoided because of entailing flicker or extinguishing of the arc. There are different possibilities:

- ECG having a low frequency output
- ECG with an superaudible operating frequency within an acoustic free window
- ECG using an operating frequency above all acoustic resonances.

Each ECG has componentries which are shown on the principle diagram. Depending on the kind of ballast additional componentries might be necessary.

The ECG with a low frequency output has an high frequency in-between circuit. This high frequency will be rectified and smoothed so that the high frequency portion is less than 10 % to avoid acoustic resonances. After that the DC-voltage is converted again to an AC voltage having a trapezoid voltage and a frequency of some 100 Hz. The advantage of such a ballast is the combination of an ECG having any high frequency and nevertheless such a low output frequency that acoustic resonances are avoided. And in spite of the low frequency no chokes having big inductivities are necessary.

Simpler are ECGs with an operating frequency in the range 20 to 50 kHz if it is possible to find a window free of acoustic resonances. Stable operation within this window must be possible when the lamps are new as well as during the whole lamp life inspite of the changes of lamp attributes.

For the time being there is only little experience for operation of HID-lamps with frequencies more than 200 kHz. This statement is valid for lamp operation as well as for the life of the ECGs itself. But operation at such a high frequency is ingenious only if there is an increased efficacy compared with operation at line frequency.

What are the difficulties when designing ECGs for operation of HID-lamps? The reason is the changing of lamp attributes between starting and operation for which the ECGs must be designed:

- Lamp ignition
- Lamp run up with overcurrent
- Lamp operation with lamps having different lamp voltages
- Extinguishing of lamps and open circuit operation
- Reignition of still warm lamps after some minutes.

ECGs must be designed for these operation modes because of two different reasons: The lamps shall be ignited carefully and operated with constant wattage, and the components of the ECG must withstand these different loads. The different loads are caused by ignition of lamps by means of overvoltage using the resonance of the 3 rd. or 5th harmonic, or because of lamp ignition at nearly any part of the half cycle so that externally controlled transistors at the end of one half cycle can lead current.

A difficult problem also is reignition of still warm lamps after some minutes of cooling down. During this time an ECG must be ready for ignition and operation.

Demonstration of an ECG for operation of Metal Halide Lamps with high power factor, constant lamp wattage and reignition of hot lamps.

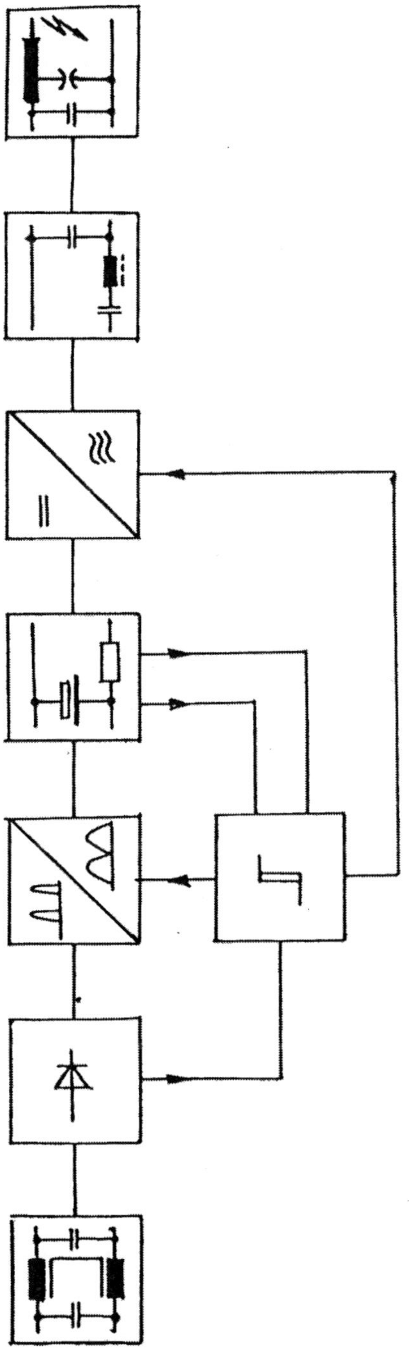

Principle diagram of an ECG for HID-lamps

NR. 73

SOME ELECTRICAL PROPERTIES OF COMPACT FLUORESCENT LAMPS AND THEIR INDUCTIVE BALLASTS.

W. Veenstra
Nederlandse Philips Bedrijven B.V.
Lighting Division
Roosendaal
The Netherlands

In literature low pressure fluorescent lamps have been described electrically by means of a small inductive element together with a resistive element 1,2). This virtual inductive character can be applied to lamps with a glass diameter of 26 and 37 mm.
From recent measurements it has been found that compact fluorescent lamps, e.g. lamps with a glass diameter of 10 mm, show an almost purely resistive behaviour.
Thus the impedance of the ballast, in series with the lamp, easily can be calculated from:

$$Z_{ballast} = \frac{-V_{la} * \cos\varphi + \sqrt{V_{la}*^2 (\cos^2\varphi - 1) + V_{supply}^2}}{I_{la}} \quad ----- (1)$$

Where $Z_{ballast}$ is the impedance of the ballast in series with a lamp.
V_{la} is the lampvoltage
$\cos\varphi$ is the power factor of the ballast
V_{supply} is the main voltage
I_{la} is the lamp current

The power dissipated in ballast ($W_{ballast}$) can be calculated from:

$$W_{ballast} = I_{la}*^2 \; Z_{ballast} * \; \cos\varphi \quad --------------- (2)$$

This calculated impedance equals within a few percent the measured impedance of the ballast in series with a lamp, with a glass diameter $\leqslant$ 15 mm.

However this calculated and measured impedance differs from the impedance of the ballast without a lamp in series.
This is due to the fact that higher harmonic wave forms arise in the circuit with a lamp.

For lamps with a glass diameter $\leqslant$ 15 mm, we found the following correlation between $Z_{ballast}$(calculated) and Z_l, the impedance of the ballast without lamp. See fig. 1.

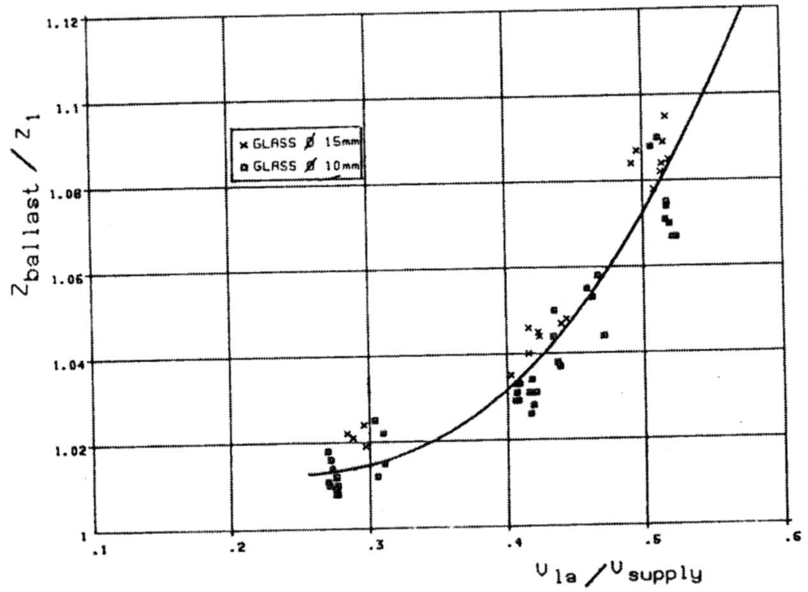

Fig.1. The ratio between the ballast impedance when the ballast is in series with a lamp and its 50 Hz sinusoidal value ($Z_{ballast}/Z_l$) as a function of the ratio between the lamp voltage and the supply voltage (V_{la}/V_{supply}).

The above mentioned relationship have proved to be rather useful in computer aided design of compact fluorescent lamps and their associated inductive ballasts.

References:

1. J.F. Waymouth : - Electric Discharge lamps.
The M.I.T. Press Cambridge, Massachusetts

2. C.H. Sturm : - Vorschaltgeräte und Schaltungen für Niederspannungs Entladungs Lampen.
Boveri & Cie , Mannheim.

Start and Operation of Compact Fluorescent Lamps at High Frequencies

Erhard Rasch
OSRAM GmbH
D-8000 München 90

Introduction

During the last five years numerable variations of fluorescent lamps have been developed. Among them are very small ones. Their dimensions are comparable with incandescent lamps but with considerable higher efficiency. It is therefore desirable to use them in a manner as simple as incandescent lamps can be used: Without additional ballasts in fixtures which are already at hand. So the ballasts, which are necessary to operate these lamps, must be integrated within the lamps.

The most effective way to reduce size and weight of a ballast is the increase of the operation frequency, facilitating the combination of lamp and ballast. Furthermore high frequency operation is advantageous for the lamp too. Lamp efficiency is increased and leads to a considerable gain of several percents.

High frequency can be generated by means of modern electronics. Normally some kind of push-pull operation is used, with transistors operating as switches.

An inductivity smoothens the rectangular shape of the switching process and limits the lamp current.

This generator for high frequency has to fulfill numerous demands for safe lamp operation:

- Save starting
- Safety against fluctuations of line voltage
- Safety against line voltage pulses
- Safety against fluctuation of temperature
- Compatibility against ambient conditions
- No critical situations when a lamp is defect or burnt out
- Low losses
- Suppression of light flickering

In order to fulfill these requirements an electronic ballast must contain further elements. These are discussed in detail.

Starting of lamps with electronic ballasts

has to be considered in more detail. At high frequencies a starting circuit with a starter switch can no longer be used. The energy which is stored in the inductance L is very low compared to 50 Hz So if we use a starter we won't get any overvoltage.

So other principles must be found. It is very easy to generate high voltage with high frequencies by resonance. Those are high enough to start the lamp without any pre-heating. In order to get sufficient lifetime of the lamp certain conditions must be met for this case: High energy in the moment of starting and precaution of any glow discharge. These conditions cannot be met with the low loading of compact lamps. In this case the electrodes of the lamp must be pre-heated before starting.

Radio interference with electronic ballasts for compact lamps

can take place either by radiation or by propagation through the leads to the line. For both cases national and international regulations are valid. In case of the radiation the magnetic stray fields of the inductivities must be suppressed by screening or some other means.

In the case of radio interference which propagates along the leads, thorough positioning of every component is necessary in order to prevent any capacitive coopling.

Realization of compact lamps presented with lamp series DULUX EL

Lifetime of compact lamps with electronic ballasts

Electronics is not a product of endless lifetime: Lamps are burning out and electronic components too. Lamps, because of consumption of emission material on the electrods. Electronics, mainly because insulation material, used in every component, becomes poorer during lifetime by the action of temperatur. An optimum in volume and prize is to make both equal under worst conditions. For this reason it is wise to integrate the ballast in the lamp and not to use adapters.

MATERIALS AND MISCELLANEOUS

75 - 80

SOLID MERCURY FILLING FOR FLUORESCENT LAMPS

W. Schuster

OSRAM GmbH, D-8000 München 90

Strict environmental protection laws require minimizing the amount of mercury in fluorescent lamps. The amount of Hg in modern single ended compact fluorescent lamps (DULUX®) can be reduced by employing recently developed pellets of a porous Fe-Cu alloy containing liquid Hg in its capillary pores. The material is wetted by Hg and does not form an amalgam so that the vapor pressure of the mercury is not reduced.

The manufacturing process of the pellets starts with an electrolysis where iron from an aqueous $FeSO_4$ solution is electrically precipitated at a continuously stirred mercury cathode. The iron-enriched mercury is mixed with a small amount of copper amalgam, covered with a thin layer of glycerol to prevent air attack, and then tempered for one hour at 250 °C.

During the tempering the CuHg is decomposed at a temperature exeeding 100 °C and an iron-copper alloy is formed. Additionally the heating causes a grain growth of the Fe crystals. Excess Hg is then filtered off and pressed away in a steel cylinder. The remaining press cake consisting of about 50% Hg and 50% Fe/Cu is hard and brittle. It is pulverized and subsequently pressed to pellets of the desired dimensions, e.g. 1.5 mm in diameter and 0.5 mm high.

Pellets offer considerable advantages over filling droplets in handling and for environmental protection:

* readily reproducible lamp filling with small Hg amounts

* reduction of environmental Hg-contamination

* easy handling of the ferromagnetic pellets

EXPONENTIAL LUMEN DEPRECIATION OF COMPACT FLUORESCENT LAMPS

B.J.Mulder and S.van Heusden
Philips Research Laboratories
5600 MD Eindhoven, The Netherlands

A study was made of the lumen depreciation of laboratory-made compact fluorescent lamps with narrow-band phosphors. The maintenance curves were found to generally fit a simple exponential curve approaching a limiting value. The decrease in light output is therefore characterized by two parameters: a rate determining exponent (k) and a preexponential factor giving the total relative lumen decrease in the limit of long periods of operation. The lumen depreciation is caused by the deposition of a black substance, mainly on the glass of the lamp wall. If the glass contains mobile alkali ions (soda lime glass, mixed alkali glass at high operating temperatures) the deposit takes the form of sharply delineated stains of irregular shape, with a remarkable fine structure of close-packed half spheres 20-30 nm across (see SEM picture).

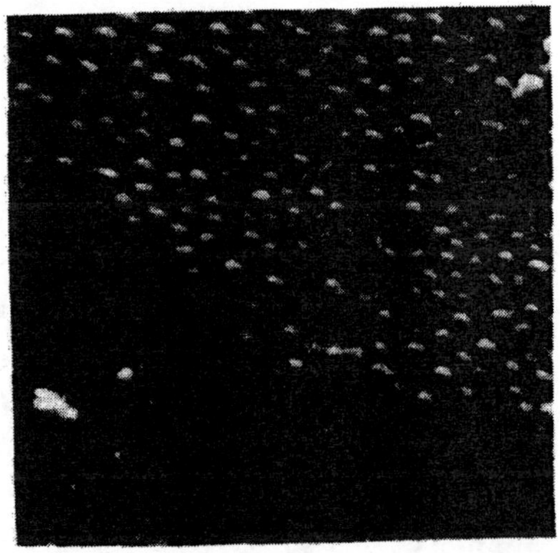

The mechanism held responsible for "exponential" lumen depreciation is the gradual conversion of colourless carbonate groups adsorbed on the phosphor grains ("CO2-ads") into a black deposit on the glass wall. The depreciation is limited, in this picture, by the amount of "CO2-ads" initially present in the lamp. To explain the exponential approach of the depreciation limit we postulate the existence in the discharge of a photochemical equilibrium between, on the one hand, "CO2-ads", strongly adsorbed at various sites and depths in the phosphor layer, and, on the other, mobile carbon monoxide and "oxygen" (see below). The exchange of CO between the various adsorption sites makes the "CO2-ads" behave as a single reservoir. CO is slowly removed from the reservoir and converted, with other ingredients (alkali oxides, mercury) into the dark deposit. The rate of removal of CO is proportional with the amount of "CO2-ads"

remaining on the phosphor, which explains exponential approach of the depreciation limit.

Transport of CO through the phosphor layer and even throughout the lamp can explain, that there is no relationship between the location of the adsorption sites (mainly on the phosphor) and the substrate of the black deposit (mainly the glass).

"Oxygen" taking part in the equilibrium between "CO2-ads" and CO + O has actually been observed in a mercury discharge, using mass spectrometry, as HgO(+)-ions, arriving at a rate of 1 in 500 Hg(+)-ions at the wall. The oxygen concentration in a mercury discharge will be governed by the formation and photodissociation of HgO and is therefore affected by such factors as the amount of HgO in the lamp (resulting mainly from water remaining in the lamp after processing), the lamp current (effect on dissociation of HgO and on transport of HgO(+)) and the location of HgO in the lamp (more or less exposed to the discharge, whether shielded or not by metallic mercury). An increase in the "oxygen" concentration in the plasma leads to a decrease in the concentration of CO (on account of the CO2-ads/CO+"O" equilibrium. This can explain that, depending on manufacturing and operating conditions, the exponential rate constant k is found to vary between 1.6 and 10 per 10,000 hours in lamps with the same phosphor, glass, diameter (10 mm) and lamp current (180 mA).

Varying the lamp current did not lead to changes in the depreciation limit, although k was affected. This supports the notion of a single reservoir of "CO2-ads" (comprising all CO2 adsorbed in the lamp), that consequently cannot be extended by increasing the intensity of the discharge.

Sodium contaminating the phosphor during lehring offers adsorption sites for CO2, thus increasing the limiting lumen loss. It also binds CO2 much more strongly than uncontaminated phosphors and therefore lowers the equilibrium concentration of CO and, hence, the rate of darkening. Finally, sodium in the glass directs the deposition of the carbon-mercury deposit to the glass wall. This may increase the effective optical density of the deposit in various ways, including the effect of repeated total reflection of light within the glass wall.

THE USE OF NIOBIUM IN HIGH PRESSURE SODIUM LAMPS

Charles I. McVey, Consulting Physicist

General Electric Company
Nela Park 5433.03
Cleveland, Ohio 44112 USA

The metal niobium has found widespread application in high pressure sodium lamps. It is used for the in-lead of the discharge chamber, radiation shields, connecting parts and supports in many designs. It is easily fabricated into useful shapes such as tubes, sheet, foil and wire. The metallurgy is well understood and with reasonable precautions the metal performs well in service. However it is necessary to understand the practical limitations of niobium in lamp processing and operation in order to obtain the desired lamp service life.

The intergranular strength of niobium is low after it has become recrystallized in lamp making operations. It is therefore necessary to take precautions to have a sufficient number of grains across the material to avoid fracture due to the stresses imparted by adjoining dissimilar materials. At the high temperatures used in making the discharge tube seals as well as in welding and brazing operations pure niobium exhibits considerable recrystallization and extensive grain growth. It is popular to use an alloy with about 1% zirconium to inhibit the grain growth properties. Even so there are practical upper limits for the time at the maximum temperature during lamp fabrication.

An important property of niobium that causes considerable difficulty is its capability to getter gaseous contaminants such as oxygen, hydrogen, and nitrogen with accompanying changes in the microstructure and in the mechanical and electrical properties of the material. Niobium dissolves these elements interstitially and when the respective solubility limit is exceeded niobium compounds are precipitated with marked changes in the materials' properties. Typically there is increasing hardness, brittleness and a bulk expansion with a commensurate increase in the electrical resistivity. Above 600 C the diffusion rate for oxygen becomes sufficient for surface oxidation to be dissolved into the body of the material. Above 800 C niobium becomes a competitor to barium getter films for oxygen scavenging of the glass outer bulb outgassing products.

This paper describes the effects of lamp processing on niobium. Welding, brazing and sealing studies are presented to indicate various errant conditions. Solutions to the typical problems are given. Included are the results of studies of oxygen contamination effects on the microstructure, the mechanical properties and the electrical resistivity of niobium and niobium-1% zirconium alloy.

Some Processing Factors Affecting the Microstructure and Properties of Translucent Alumina Tubes

Zheng Shaoxian
Shanghai Institute of Ceramics, Academia Sinica

When the ammonium alum is used as a precusor of the starting alumina powder for manufacturing translucent alumina ceramic tubes, its crystal habit has a great influence on the morphology and particle size distribution of the alumina powders, the parameters of the alum preparation observed in this investigation involved the temperature during crystallization, the concentration of mother liquid and the additions.Two typical alumina powders derived from alum crystallized under different conditions but calcined at the same temperature showed quite different characteristics. Electron microscopy study revealed that one powder consisted of very fine and well dispersed particles and contained a small amount of r-phase, the ceramic specimens prepared using this powder achieved high degree of translucenty and their microstructure was pore-free and homogeneous, thus the lamp efficacy is high. While the other alumina powder was composed mainly of coarse and high agglomerated particles which were very difficult to dispers even ofter ball-milling for a number of hour, and no r-phase was found. Ceramic specimens from the powder showed a duplex microstructure with many small pores trapped inside the grains and, as a result, they were nearly opaque, then lamp efficacy is low. Judging from these experimental results of the latter powder is mainly caused by its agglomerate structure, thus resulting in a green body with an uneven microstructure containing a large amount of big sized pores which are very difficult to be finally eliminated during sintering.

Electron microprobe analysis has detected a few spots containing both sodium and calcium impurities. As a consequence, we suppose the calcium containing phase to be detrimantal impurities in the translucent ceramics tube which decrease the resistance of the alumina ceramics aginst sodium.

Translucent alumina tube used for high pressure sodium lamp

Gao Long Qiao, Li Fa

BEIJING VACUUM ELECTRON DEVICES RESEARCH INSTITUTE
P.O.Box 749, Beijing

A kind of translucent alumina tube of high total light transmittance (94.1-96.0 %), low sintering temperature(1680-1720°C) and short sintering period (4 hours) have been developed successfully. Several processing factors affecting the ceramic microstructure and properties have been investigated also. Alumina powder is derived from domestic ammonium alum. The powder was used which was nominally 99.9 % pure with no other X-ray detectable phases than α-Al_2O_3 and was characterized with following results:

impurities (by emission spectroscopy)

Sn	Ag	Ga	Zn	Cu	Ni	Fe	Mn	Ca	K	S	P	Si	Mg	Na	B
content (PPM)															
3	4	110	12	2	0.5	23	0.2	12	30	11	2	77	28	4	5

specific surface area 5.04 m^2/g

average particle size 0.32 μm

Particularly point out that our work adopt composite additive (codoping) first and the result is exciting. The composite additive formulate following:

Y_2O_3	0.05-0.3	wt%
La_2O_3	o.05-0.5	wt%
MgO	0.01-0.3	wt%
ZnO	0.01-0.3	wt%

The fracture surface of the sample was analyzed with SEM, XPS and ion probe microanalyzer for research of fast sintering mechanism, which showed that a solid solution model is a more possible mechanism.

SEM photograph shows an uniform, homogeneous and fine grain microstructure, and rupture occurs along grain boundary mainly. There are existence of a few phases enriched Y_2O_3 at grain boundary. Both XPS and ion probe microanalyzer determine that there is no La_2O_3 or a very small amount La_2O_3 in the alumina tubes. Therefore benefit of La_2O_3 added as regulating the index of refraction that is traditional opinion may be suspicious. ZnO additive added has obtained obvious functions of lowering sintering temperature and promoting fast sintering until near theoretical density.

The characteristics of the resultant alumina tube obtained are also presented as follows:

tolerences of tube (for 400 W)

- oblateness $\leq$ 0.067 mm
- bending $\leq$ 0.068 mm
- wall thickness $\pm$ 0.089 mm
- density 3.987 g/cm^3
- thermal shock resistance (thrown to water) $\geq 240\,^\circ C$
- thermal expansion coefficient (29-900°C) $8.07 \times 10^{-6}\,^\circ C^{-1}$
- average grain size 11.0 μm
- total light transmittance % 94.1-96.0

References

1. R.L.Coble J. Amer. Ceram. Soc. 41(2) 55 (1958)
2. R.L.Coble J. Appl. Phys. 32 (5) 787-793 (1961)
3. R.T.Temper et al J. Amer. Ceram. Soc. 57, 421 (1974)
4. T.G.Langdon et al J. Mater. Sci., 11, 317 (1976)
5. U.S. Patent 4,285,732

CATHODOLUMINESCENCE AS A LIGHT SOURCE

AUTHOR:- S.A.R. RIGDEN

ADVANCED DEVELOPMENT LABORATORY,

OSRAM (GEC) LIMITED,

EAST LANE, WEMBLEY, MIDDLESEX,

HA9 7PG, ENGLAND.

A re-assessment has been made of the possibility of utilising cathodoluminescence when designing a practical light source as an alternative to discharge and incandescent lamps. The investigation was initiated as a result of current advances in cathode ray tube design and miniaturisation of electronics from solid state technology. The development of a novel flood gun is described, which is capable of uniformly illuminating the whole surface of all bulbs likely to be required as light sources. The investigation includes a survey of a considerable range of phosphors in which the separate effects on light output due to electron bombardment and heating will be discussed. A solid state circuit has been designed for lamp operation and will be described.

POSTDEADLINE PAPERS

81 -

Metal Oxide Containing High Intensity Discharge Lamp

W.P. Lapatovich; GTE Laboratories, Inc., W.M. Keeffe, N. Brates,
R.W. Liebermann, and J. Maya; GTE Lighting Products,
100 Endicott St., Danvers, MA 01923 U.S.A.

We have investigated a novel high pressure electric discharge lamp which exhibits desired properties of high efficacy, good color rendering, warm color temperature, and reasonable life. These desired properties are attained by a novel prescription of the fill to include strong atomic and molecular emitters.

The molecular additives which augment the radiative output are chosen to have high volatility, fragmentation products which emit useful radiation, stability of the fragment at arc temperatures, and chemical compatibility with arc tube components. An example of such a molecule is $NbOCl_3$. The desired radiating fragment, NbO, is stable at elevated temperatures (ground state dissociation energy D=7.8eV) and has a spectrum rich in the red where most metal halide lamps are deficient.

The desired radiating fragments, Nb and NbO, are easily introduced into the vapor phase via collisional dissociation of the more volatile parent molecules in the discharge. These fragments are excited further by electron impact, producing excited states which radiate the useful visible radiation (see Fig. 1).

Thermodynamics calculations were performed to predict fill species composition in the vapor phase and above the melt. These compositions were then used to predict concentrations at elevated temperatures and to simulate plasma-materials interactions with the electrodes and the vessel wall. The predicted modification of the vapor composition was used as a guide in tailoring lamp fills and in choosing fabrication materials. The lamps investigated consisted of a vitreous silica arc tube filled with Hg, Ar, and other chemical components designed to produce $NbOCl_3$ in situ in the electric discharge.

The desired radiating fragments Nb, and NbO are produced by subsequent collisional dissociation in the discharge. These fragments are excited further by electron impact, producing excited states which radiate the useful visible radiation (see Fig. 1).

Both electrodeless (915MHz) and electroded (60Hz) lamps were investigated (2.4 and 4.0cm^3 respectively) with the $NbOCl_3$ fill. Typical results are summarized in Table 1 below:

Table 1

Parameter	175 W Electroded Lamp	Electrodeless Lamp
Wall loading (W/cm^2)	27.7	27.3
Efficacy (lm/W)	72.9	85.1
General CRI	94.2	to be measured
Correlated color temperature (K)	6263	

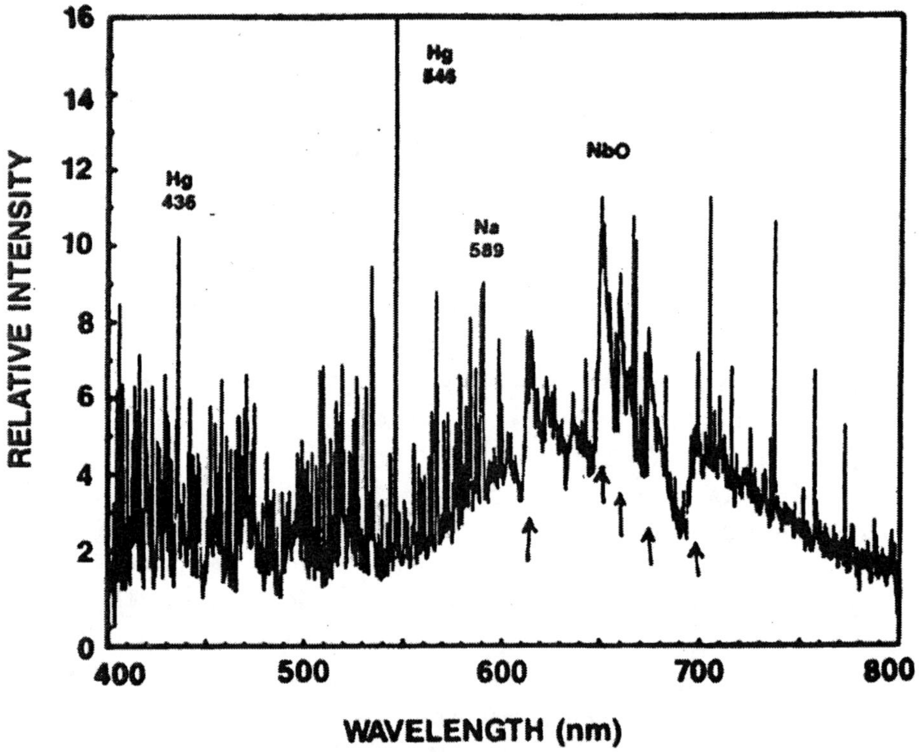

Figure 1. Energy corrected emission spectrum from an electrodeless $NbOCl_3$ arc lamp.

*This work was partially supported by the U.S. Department of Energy through a subcontract from Lawrence Berkeley Laboratory.

AUTHORS AND CO-AUTHORS